W9-CMM-989

The IMA Volumes in Mathematics and its Applications

Volume 61

Series Editors
Avner Friedman   Willard Miller, Jr.

Institute for Mathematics and its Applications
IMA

The **Institute for Mathematics and its Applications** was established by a grant from the National Science Foundation to the University of Minnesota in 1982. The IMA seeks to encourage the development and study of fresh mathematical concepts and questions of concern to the other sciences by bringing together mathematicians and scientists from diverse fields in an atmosphere that will stimulate discussion and collaboration.

The IMA Volumes are intended to involve the broader scientific community in this process.

Avner Friedman, Director
Willard Miller, Jr., Associate Director

* * * * * * * * * *

IMA ANNUAL PROGRAMS

1982–1983 **Statistical and Continuum Approaches to Phase Transition**
1983–1984 **Mathematical Models for the Economics of Decentralized Resource Allocation**
1984–1985 **Continuum Physics and Partial Differential Equations**
1985–1986 **Stochastic Differential Equations and Their Applications**
1986–1987 **Scientific Computation**
1987–1988 **Applied Combinatorics**
1988–1989 **Nonlinear Waves**
1989–1990 **Dynamical Systems and Their Applications**
1990–1991 **Phase Transitions and Free Boundaries**
1991–1992 **Applied Linear Algebra**
1992–1993 **Control Theory and its Applications**
1993–1994 **Emerging Applications of Probability**

IMA SUMMER PROGRAMS

1987 **Robotics**
1988 **Signal Processing**
1989 **Robustness, Diagnostics, Computing and Graphics in Statistics**
1990 **Radar and Sonar**
1990 **Time Series**
1991 **Semiconductors**
1992 **Environmental Studies: Mathematical, Computational, and Statistical Analysis**

* * * * * * * * * *

SPRINGER LECTURE NOTES FROM THE IMA:

The Mathematics and Physics of Disordered Media
Editors: Barry Hughes and Barry Ninham
(Lecture Notes in Math., Volume 1035, 1983)

Orienting Polymers
Editor: J.L. Ericksen
(Lecture Notes in Math., Volume 1063, 1984)

New Perspectives in Thermodynamics
Editor: James Serrin
(Springer-Verlag, 1986)

Models of Economic Dynamics
Editor: Hugo Sonnenschein
(Lecture Notes in Econ., Volume 264, 1986)

Robert A. Brown Stephen H. Davis
Editors

Free Boundaries in Viscous Flows

With 46 Illustrations

Springer-Verlag
New York Berlin Heidelberg London Paris
Tokyo Hong Kong Barcelona Budapest

Robert A. Brown
Department of Chemical Engineering
Massachusetts Institute of Technology
Room 66-353
Cambridge, MA 02139 USA

Stephen H. Davis
Department of Engineering Science
and Applied Mathematics
Northwestern University
Evanston, IL 60208 USA

Series Editors:
Avner Friedman
Willard Miller, Jr.
Institute for Mathematics and its
Applications
University of Minnesota
Minneapolis, MN 55455 USA

Mathematics Subject Classifications (1991): 76Exx, 35K22, 35B32, 76B45, 76D05

Library of Congress Cataloging-in-Publication Data
Free boundaries in viscous flows / Robert A. Brown, Stephen H. Davis, editors.
p. cm. – (The IMA volumes in mathematics and its applications : v. 61)
Includes bibliographical references.
ISBN 0-387-94253-X (New York). – ISBN 0-540-94253-X (Berlin)
1. Viscous flow—Mathematical models—Congresses. I. Brown, Robert A. (Robert Arthur). II. Davis, Stephen H., 1939–. III. Series.
QA929.F73 1994 93-50711
532′.0533—dc20

Printed on acid-free paper.

Production managed by Hal Henglein; manufacturing supervised by Vincent Scelta.
Camera-ready copy prepared by the IMA.
Printed and bound by Edwards Brothers, Inc., Ann Arbor, MI.
Printed in the United States of America.

9 8 7 6 5 4 3 2 1

ISBN 0-387-94253-X Springer-Verlag New York Berlin Heidelberg
ISBN 3-540-94253-X Springer-Verlag Berlin Heidelberg New York

The IMA Volumes
in Mathematics and its Applications

Current Volumes:

Volume 1: Homogenization and Effective Moduli of Materials and Media
Editors: Jerry Ericksen, David Kinderlehrer, Robert Kohn, J.-L. Lions

Volume 2: Oscillation Theory, Computation, and Methods of Compensated Compactness
Editors: Constantine Dafermos, Jerry Ericksen,
David Kinderlehrer, Marshall Slemrod

Volume 3: Metastability and Incompletely Posed Problems
Editors: Stuart Antman, Jerry Ericksen, David Kinderlehrer, Ingo Muller

Volume 4: Dynamical Problems in Continuum Physics
Editors: Jerry Bona, Constantine Dafermos, Jerry Ericksen,
David Kinderlehrer

Volume 5: Theory and Applications of Liquid Crystals
Editors: Jerry Ericksen and David Kinderlehrer

Volume 6: Amorphous Polymers and Non-Newtonian Fluids
Editors: Constantine Dafermos, Jerry Ericksen, David Kinderlehrer

Volume 7: Random Media
Editor: George Papanicolaou

Volume 8: Percolation Theory and Ergodic Theory of Infinite Particle Systems
Editor: Harry Kesten

Volume 9: Hydrodynamic Behavior and Interacting Particle Systems
Editor: George Papanicolaou

Volume 10: Stochastic Differential Systems, Stochastic Control Theory and Applications
Editors: Wendell Fleming and Pierre-Louis Lions

Volume 11: Numerical Simulation in Oil Recovery
Editor: Mary Fanett Wheeler

Volume 12: Computational Fluid Dynamics and Reacting Gas Flows
Editors: Bjorn Engquist, M. Luskin, Andrew Majda

Volume 13: Numerical Algorithms for Parallel Computer Architectures
Editor: Martin H. Schultz

Volume 14: Mathematical Aspects of Scientific Software
Editor: J.R. Rice

Volume 15: Mathematical Frontiers in Computational Chemical Physics
Editor: D. Truhlar

Volume 16: Mathematics in Industrial Problems
by Avner Friedman

Volume 17: Applications of Combinatorics and Graph Theory to the Biological and Social Sciences
Editor: Fred Roberts

Volume 18: q-Series and Partitions
Editor: Dennis Stanton

Volume 19: Invariant Theory and Tableaux
Editor: Dennis Stanton

Volume 20: Coding Theory and Design Theory Part I: Coding Theory
Editor: Dijen Ray-Chaudhuri

Volume 21: Coding Theory and Design Theory Part II: Design Theory
Editor: Dijen Ray-Chaudhuri

Volume 22: Signal Processing: Part I - Signal Processing Theory
Editors: L. Auslander, F.A. Grünbaum, J.W. Helton, T. Kailath, P. Khargonekar and S. Mitter

Volume 23: Signal Processing: Part II - Control Theory and Applications of Signal Processing
Editors: L. Auslander, F.A. Grünbaum, J.W. Helton, T. Kailath, P. Khargonekar and S. Mitter

Volume 24: Mathematics in Industrial Problems, Part 2
by Avner Friedman

Volume 25: Solitons in Physics, Mathematics, and Nonlinear Optics
Editors: Peter J. Olver and David H. Sattinger

Volume 26: Two Phase Flows and Waves
Editors: Daniel D. Joseph and David G. Schaeffer

Volume 27: Nonlinear Evolution Equations that Change Type
Editors: Barbara Lee Keyfitz and Michael Shearer

Volume 28: Computer Aided Proofs in Analysis
Editors: Kenneth Meyer and Dieter Schmidt

Volume 29: Multidimensional Hyperbolic Problems and Computations
Editors: Andrew Majda and Jim Glimm

Volume 30: Microlocal Analysis and Nonlinear Waves
Editors: Michael Beals, R. Melrose and J. Rauch

Volume 31: Mathematics in Industrial Problems, Part 3
by Avner Friedman

Volume 32: Radar and Sonar, Part I
by Richard Blahut, Willard Miller, Jr. and Calvin Wilcox

Volume 33: Directions in Robust Statistics and Diagnostics: Part I
Editors: Werner A. Stahel and Sanford Weisberg

Volume 34: Directions in Robust Statistics and Diagnostics: Part II
Editors: Werner A. Stahel and Sanford Weisberg

Volume 35: Dynamical Issues in Combustion Theory
Editors: P. Fife, A. Liñán and F.A. Williams

Volume 36: Computing and Graphics in Statistics
Editors: Andreas Buja and Paul Tukey

Volume 37: Patterns and Dynamics in Reactive Media
Editors: Harry Swinney, Gus Aris and Don Aronson

Volume 38: Mathematics in Industrial Problems, Part 4
by Avner Friedman

Volume 39: Radar and Sonar, Part II
Editors: F. Alberto Grünbaum, Marvin Bernfeld and Richard E. Blahut

Volume 40: Nonlinear Phenomena in Atmospheric and Oceanic Sciences
Editors: George F. Carnevale and Raymond T. Pierrehumbert

Volume 41: Chaotic Processes in the Geological Sciences
Editor: David A. Yuen

Volume 42: Partial Differential Equations with Minimal Smoothness and Applications
Editors: B. Dahlberg, E. Fabes, R. Fefferman, D. Jerison, C. Kenig, and J. Pipher

Volume 43: On the Evolution of Phase Boundaries
Editors: Morton E. Gurtin and Geoffrey B. McFadden

Volume 44: Twist Mappings and Their Applications
Editor: Richard McGehee and Kenneth R. Meyer

Volume 45: New Directions in Time Series Analysis, Part I
Editors: David Brillinger, Peter Caines, John Geweke, Emanuel Parzen, Murray Rosenblatt, and Murad S. Taqqu

Volume 46: New Directions in Time Series Analysis, Part II
Editors: David Brillinger, Peter Caines, John Geweke, Emanuel Parzen, Murray Rosenblatt, and Murad S. Taqqu

Volume 47: Degenerate Diffusions
Editors: Wei-Ming Ni, L.A. Peletier, and J.-L. Vazquez

Volume 48: Linear Algebra, Markov Chains and Queueing Models
Editors: Carl D. Meyer and Robert J. Plemmons

Volume 49: Mathematics in Industrial Problems, Part 5
by Avner Friedman

Volume 50: Combinatorial and Graph-Theoretical Problems in Linear Algebra
Editors: Richard A. Brualdi, Shmuel Friedland, and Victor Klee

Volume 51: Statistical Thermodynamics and Differential Geometry of Microstructured Materials
Editors: H. Ted Davis and Johannes C.C. Nitsche

Volume 52: Shock Induced Transitions and Phase Structures in General Media
Editors: J.E. Dunn, Roger Fosdick, and Marshall Slemrod

Volume 53: Variational and Free Boundary Problems
Editors: Avner Friedman and Joel Spruck

Volume 54: Microstructure and Phase Transition
Editors: David Kinderlehrer, Richard James, Mitchell Luskin and Jerry L. Ericksen

Volume 55: Turbulence in Fluid Flows: A Dynamical Systems Approach
Editors: George R. Sell, Ciprian Foias, and Roger Temam

Volume 56: Graph Theory and Sparse Matrix Computation
Editors: Alan George, John R. Gilbert, and Joseph W.H. Liu

Volume 57: Mathematics in Industrial Problems, Part 6
by Avner Friedman

Volume 58: Semiconductors, Part I
Editors: W.M. Coughran, Jr., Julian Cole, Peter Lloyd, and Jacob White

Volume 59: Semiconductors, Part II
Editors: W.M. Coughran, Jr., Julian Cole, Peter Lloyd, and Jacob White

Volume 60: Recent Advances in Iterative Methods
Editors: Gene Golub, Anne Greenbaum, and Mitchell Luskin

Volume 61: Free Boundaries in Viscous Flows
Editors: Robert A. Brown and Stephen H. Davis

Forthcoming Volumes:

1989-1990: *Dynamical Systems and Their Applications*

Dynamical Theories of Turbulence in Fluid Flows

1991-1992: *Applied Linear Algebra*

Linear Algebra for Signal Processing

Linear Algebra for Control Theory

Summer Program *Environmental Studies*

Environmental Studies

FOREWORD

This IMA Volume in Mathematics and its Applications

FREE BOUNDARIES IN VISCOUS FLOWS

is based on the proceedings of a workshop which was an integral part of the 1990-91 IMA program on "Phase Transitions and Free Boundaries." The workshop addressed the mathematical treatment of the dynamical and instability phenomena in the interaction of viscous flows with liquid/solid interfaces and solidification fronts; it brought together mathematicians and materials scientists with interests in these problems. We thank R. Fosdick, M.E. Gurtin, W.-M. Ni, and L.A. Peletier for organizing the year-long program.

We are particularly grateful to Robert A. Brown and Stephen H. Davis for organizing this workshop and editing the proceedings.

We also take this opportunity to thank the National Science Foundation, whose financial support made the workshop possible.

Avner Friedman
Willard Miller, Jr.

PREFACE

It is increasingly the case that models of natural phenomena and materials processing systems involve viscous flows with free surfaces. These free boundaries are interfaces of the fluid with either second immiscible fluids or else deformable solid boundaries. The deformation can be due to mechanical displacement or, as is the case here, due to phase transformation; the solid can melt or freeze.

The need for the analysis of viscous, free-surface problems has put new emphasis on the mathematical understanding of the structure of these problems. For example, it has become clear that the subtle physicochemical details at moving and free contact lines between fluids and solid surfaces have dramatic effects on the well-posedness of the boundary-value problems that arise. Moreover, such mathematical understanding is necessary for the construction of the convergent and stable numerical methods for solution.

The Symposium in March, 1991, entitled "Free Boundaries in Viscous Flows," was sponsored by the Institute for Mathematics and its Applications (IMA) of the University of Minnesota at Minneapolis in order to bring together 40 leading researchers interested in such problems. The participants ranged from mathematical analysts to material scientists, all with common interests in free-boundary problems. This volume of the IMA Proceedings highlights the range of subjects discussed at this meeting. The paper by Young gives an overview of the mathematical description of viscous free-surface flows. The current understanding of mathematical issues that arise in these models is addressed in the paper by Pritchard, Saavedra, Scott, and Tavener. Higdon and Schnepper discuss a high-order-accuracy boundary-integral method for solution of viscous free-surface flows. Examples of the mathematical analysis of particular flows are described in the overview by Smith of long-wave instabilities in viscous-film flows, by Pukhnachov's analysis of long-wave instabilities leading to Marangoni convection, and by Coriell, Murray, McFadden, and Leonartz's description of the interaction of convection with morphological stability during directional solidification.

The organizers are indebted to Avner Friedman, the Director of the IMA, for nucleating the idea of the meeting and for providing a warm and friendly setting for an open exchange of ideas. We also are indebted to Ms. Kaye Smith for her efforts in organizing this volume; without her, the volume would not exist.

Robert A. Brown
Stephen H. Davis

CONTENTS

MATHEMATICAL DESCRIPTION OF VISCOUS FREE SURFACE FLOWS

GERALD W. YOUNG *

Abstract. Mathematical formulations for three distinct systems which involve viscous free surface flows are presented. The first model describes boundary conditions posed at a moving contact line formed at the tri-junction of a gas/liquid/solid system. In particular the unification of mass flux conditions is highlighted. The second and third models are solidification systems. Formulations of both models examine the use of asymptotic methods to isolate a mathematical description of the free solidifying surface. In one case evolution equations for solidifying fronts are developed. These equations allow one to investigate micro scale morphology of the interface. In the other model, the influence of macroscopic heat, mass, and momentum transport on the solidifying front is examined. A reduced model is developed which is simpler for analysis yet retains the relevant physics.

Key words. contact line, evolution equation, free surface, solidification, viscous flow

AMS(MOS) subject classifications. 35C20, 76D99

1. Introduction. The objective of this discussion is to formulate mathematical models for three systems involving viscous free surface flows. The first model concerns the development of boundary conditions at a contact line. The second and third models are solidification systems. One concerns an analysis of the microscopic behavior at the solidification front while the other involves a macroscopic investigation of heat, mass, and momentum transport on the front. The analysis and solution of these models will not be explored in much detail. However, we shall examine the identification of several key asymptotic limits in the solidification systems which may be exploited in an asymptotic solution methodology. Asymptotics, in this context, will be the study of the local behavior of functions (in this case partial differential equations resulting from a continuum description of the system) in the neighborhood of a point (limiting cases in parameter space). Such a local study is necessary since the mathematical models are so complex that exact analytical treatment is intractable. Hence, the goal of investigating limiting cases is to reduce the complexity of the model but retain as much of the physics as possible. In the models developed in this discussion, we shall focus on an asymptotic strategy by which one "avoids" doing a free-boundary problem. Rather one isolates a mathematical description of the free surface, from which its location, dynamics and stability can be investigated. This isolated mathematical description may be a single evolution equation or a system of equations. In either case, this reduced system is simpler for analytical or numerical investigation.

When using the reduced model, one must be aware of its purpose when considering how well it represents the original system. Often times asymptotic models, by intent, focus on a few, maybe even just one, aspects of the full system. For example, fundamental physics, such as the dynamics of moving contact lines, may be poorly understood. Hence, a model concentrating solely on the behavior of the contact line

* Department of Mathematical Sciences, The University of Akron, Akron, Ohio 44325-4002.
This work is supported by NSF Grant DMS-89-57534 and by the Institute for Mathematics and its Applications with funds provided by the National Science Foundation.

may be appropriate for elucidating the physics. In other cases the physical mechanisms present in a system may be understood in isolation from one another, such as morphological versus convective instabilities in a solidification system, but not well understood when both are present simultaneously. In these instances, where the fundamental questions involve basic science issues, reduced models may give insight into the behavior of the full system.

In terms of quantitative predictions from the reduced model, it is clear that asymptotic analyses are only quantitatively valid when one is faithful to the limits. 'How small is small enough for quantitative accuracy' can only be answered through comparison with exact (which are usually unavailable) or numerical solutions. Often times, the region of validity is not too restrictive, as we shall show in one example. On the other hand, there are situations where the asymptotic limits impose such severe constraints on the model, that no real system satisfies all the restrictions. In this case, the best one can say is that the behavior of the reduced model is suggestive of the behavior outside the region of validity. Such may be the situation for long-wavelength evolution equations of solidification fronts.

2a. Contact-line boundary conditions. The motion of a contact line formed by, say, a liquid, a gas and a solid is part of many systems of scientific and engineering interest. These include coating and spreading, mutual displacement in porous media and the opening of dry patches on heated plates.

Consider a liquid/gas/solid system as shown in Fig. 1. Here there is a fluid interface at $y = h(x, z, t)$ with a contact line at $y = 0$, $x = A(z, t)$. In general, contact-line systems are free boundary problems so that both h and A are *a priori* unknown. In order to develop well-posed boundary value problems in such cases, boundary conditions at the contact line are required.

As reviewed in [5], there are generally two conditions defined: i) condition of contact and ii) condition of contact angle θ depending on contact-line speed U_{CL}. The first of these defines the intersection of the interface with the solid. Here

$$h[A(z,t), z, t] = 0. \tag{2.1}$$

The second of these gives the local slope of h at the contact line in terms of θ, where

$$\theta = G(U_{CL}). \tag{2.2}$$

The contact-line speed U_{CL}, measured relative to the solid in the direction normal to the contact line, is taken positive if liquid displaces gas and negative if gas displaces liquid. The function G relates the thermodynamic contact angle θ to U_{CL}. See [9] for a detailed discussion.

Beside the specification of (2.1) and that stemming from (2.2), other conditions are required. Greenspan [11] and Greenspan and McCay [12] derive a fourth-order partial differential equation for h governing the spreading of an axisymmetric liquid drop on a plate. One thus requires conditions to fix the four integration constants plus the contact-line position. Forms (2.1) and (2.2) give two conditions at the contact line. Since the drop spreads axisymmetrically, they assume center-line symmetry and constant liquid volume to determine the remaining conditions. Analogous procedures have been used by Hocking [13], Lacey [16] and Dussan V. and Chow [10] where *a*

All of section 2 is reprinted from [29].

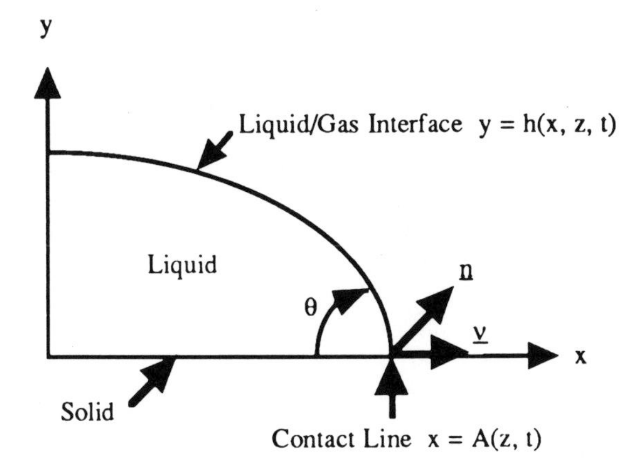

a) Side View

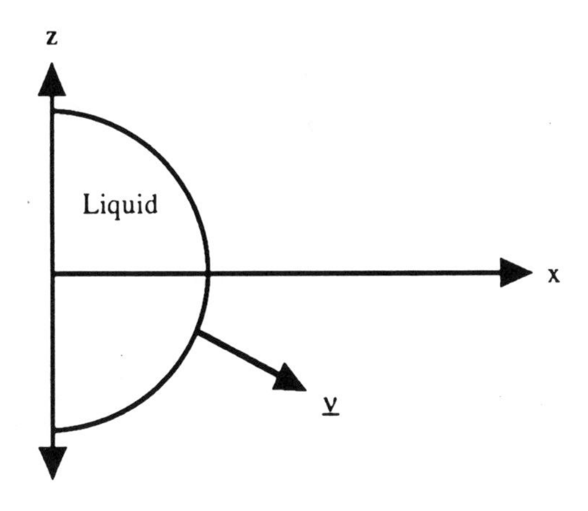

b) Plan View

FIG. 1. *Sketch of a gas/liquid/solid system possessing a contact line.* $\underline{n}$ *is the normal to the interface and* $\underline{\nu}$ *is the normal to the contact line in the plane of the solid.*

priori symmetries are known. However, if such symmetries are absent, one must pose a more primitive condition. This is required in problems of instability in systems containing contact lines since the preferred mode of instability is to be determined. Rivulet instabilities (Davis [4], Weiland and Davis [27]) are in this class. Young [28] develops an evolution equation for h using lubrication theory for thin flat rivulets subject to long-wave instabilities. He finds that the evolution equation is fourth order in space and six conditions are required to determine the four integration constants and two contact-line positions. In special situations a boundedness condition can be posed in lieu of one boundary condition. Hocking [13] does this and shows its equivalence to a volume constraint. Lacey [16] poses a similar condition and shows its equivalence to local mass conservation at the contact line.

The purpose of the present section is to formulate explicitly a new primitive boundary condition to be satisfied at a contact line. The new condition is one of zero-mass-flux at the contact line. This condition together with the contact condition and the contact-angle-varying-with-speed condition unifies the prescriptions for any geometry. Its equivalence in special cases to symmetry and volume conservation is shown. Further, a new kinematic condition, equivalent to the zero-mass-flux condition, is obtained. This latter form is more easily incorporated in numerical simulations.

2b. Zero-mass flux condition. Let us consider the local geometry of a moving contact line as shown in Fig. 1. Here $\underline{v} = (u, v, w)$ is the velocity in a Cartesian system (x, y, z) and $\underline{\nu}$ is the outward normal to the contact line in the plane of the solid plate. If the liquid is pure and the system is isothermal, then the governing equations are the constant-density Navier-Stokes and continuity, the latter being

$$u_x + v_y + w_z = 0. \tag{2.3}$$

These are subject to boundary conditions on the liquid-gas interface at $y = h(x, z, t)$. These are the normal and tangential stress balances plus the kinematic condition

$$v = h_t + uh_x + wh_z, \quad y = h. \tag{2.4}$$

There are conditions on the liquid-solid interface, viz. zero penetration

$$v = 0, \quad y = 0 \tag{2.5}$$

plus, say, conditions of apparent slip,

$$f_i(u, w) = 0, \quad i = 1, 2 \quad y = 0 \tag{2.6}$$

consistent with the conclusion of Dussan V. and Davis [8] from kinematics that a moving contact line and the no slip condition result in a nonintegrable singularity at the contact line. If conditions (2.6) yield perfect slip at the contact line, then the singularity is removed.

Finally, there are contact-line conditions governing contact, equation (2.1), and governing the contact angle

$$\underline{\nu} \cdot \nabla h[A(z, t), z, t] = -\tan\theta \tag{2.7}$$

where θ is given by eqn. (2.2).

The additional condition is *implicit* in all the models of moving contact lines considered to date, namely that no liquid escapes through the contact line. We state this condition as the vanishing of the liquid mass flux through the contact line. If the density of the liquid is constant, then the horizontal volume flux $\underline{Q}$,

$$\underline{Q} = \underline{i} \int_0^h u \, dy + \underline{k} \int_0^h w \, dy \tag{2.8}$$

should satisfy the no leakage condition, viz.

$$\underline{Q} \cdot \underline{\nu} = 0 \text{ at } y = 0, \; x = A. \tag{2.9}$$

This is a *primitive boundary condition* at each line. It implies that the kinematic condition (2.4) applies on the interface everywhere including the contact line and that the contact line is a boundary of the system.

To see this differentiate equation (2.1) with respect to z,

$$h_z = -h_x A_z \text{ at } x = A, \; y = 0 \tag{2.10}$$

Now differentiate eqn. (2.1) with respect to t,

$$h_t = -h_x A_t \text{ at } x = A, \; y = 0. \tag{2.11}$$

Now, evaluate the kinematic condition (2.4) at the contact line, use eqn. (2.5) and forms (2.10) and (2.11) to obtain

$$h_x(-A_t + u - wA_z) = 0 \tag{2.12}$$

or, if $h_x \neq 0$, then

$$u = A_t + wA_z \text{ at } x = A, \; y = 0. \tag{2.13}$$

Eqn. (2.13) is a kinematic condition at the contact line in the plane of the solid. It gives the balance between the speed U_{CL} of propagation of the contact line

$$U_{CL} = A_t(1 + A_z^2)^{-\frac{1}{2}} \tag{2.14}$$

and the normal component $\underline{v} \cdot \underline{\nu}$ of the fluid at the contact line.

Let us take an arbitrary surface S within the fluid (as shown in Fig. 2) which is oriented perpendicular to the plate and parallel in shape to the contact line $x = A$. The mass flux through this area is given by

$$Q_S = \int_{z_1}^{z_2} \underline{Q} \cdot \underline{\nu} \, dz. \tag{2.15}$$

Now let the surface approach the contact line. Define

$$\tilde{Q}_s = \lim_{S \to CL} Q_S = \int_{z_1}^{z_2} [\lim_{S \to CL} \underline{Q} \cdot \underline{\nu}] \, dz, \tag{2.16}$$

which is the mass flux through the contact line. However, this mass flux can also be defined as

$$\begin{aligned} \tilde{Q}_{CL} \;=\; & \textstyle\int_{z_1}^{z_2} [\text{speed of the fluid normal to the contact line} \\ & - \text{speed of the contact line}] \, dz. \end{aligned} \tag{2.17}$$

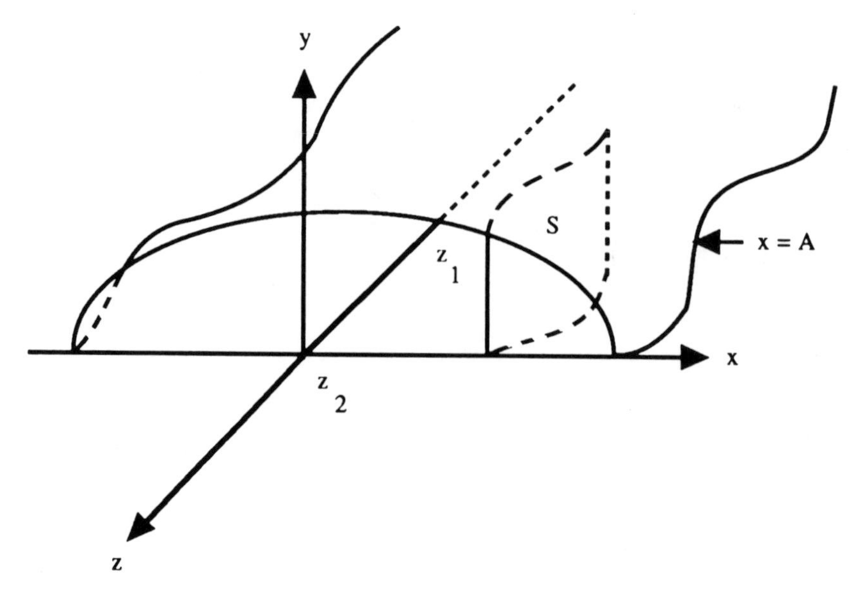

FIG. 2. *The area S is parallel in shape to the contact line $x = A$.*

From relation (2.14) we find that

$$\tilde{Q}_{CL} = \int_{z_1}^{z_2} [(u, w) \cdot \underline{\nu} - U_{CL}]\, dz$$

$$= \int_{z_1}^{z_2} [(u, w) \cdot (1, -A_z) - A_t](1 + A_z^2)^{-\frac{1}{2}}\, dz$$

$$= \int_{z_1}^{z_2} [-A_t - wA_z + u](1 + A_z^2)^{-\frac{1}{2}}\, dz\,. \tag{2.18}$$

Since eqns. (2.16) and (2.18) both give the mass flux through the contact line, we have

$$\lim_{S \to CL} Q_S = \tilde{Q}_{CL}. \tag{2.19}$$

But as stated previously, there is no mass flux through any portion of a contact line because contact lines define part of the fluid boundary. Therefore, both expressions (2.16) and (2.18) must be zero regardless of the chosen portion (z_1, z_2) of the contact line. This requires the integrands of both to be equal so that

$$\lim_{S \to CL} \underline{Q} \cdot \underline{\nu} = 0 \tag{2.20}$$

and

$$-A_t - wA_z + u = 0. \tag{2.21}$$

But relation (2.20) is just relation (2.9) and relation (2.21) is exactly relation (2.13). Thus, the zero mass flux condition guarantees that (2.13) is satisfied. In turn, (2.13) guarantees that

$$v(A(z, t), y, z) = 0, \tag{2.22}$$

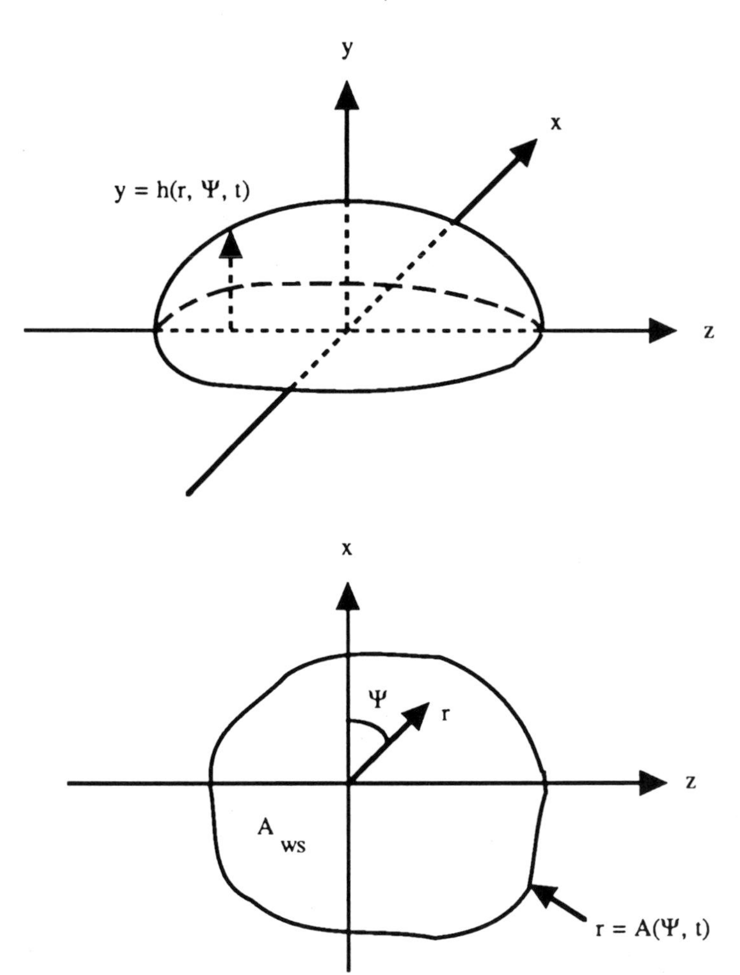

FIG. 3. *Sketch of a drop of liquid on a smooth solid surface. The wetted area A_{ws} of the solid is given by $r = A(\Psi, t)$.*

since by evaluating condition (2.4) at $x = A(z,t)$ we obtain

$$h_t + uh_x + wh_z = v(A(z,t), y, z) = h_x(-A_t - wA_z + u), \tag{2.23}$$

using relations (2.10) and (2.11).

2c. Equivalences. As mentioned previously, boundedness conditions and volume constraints have previously been posed in lieu of the zero-mass-flux condition in forms (2.9) or (2.13). Greenspan and McCay [12] state a condition similar to (2.13). They then abandon it and revert to a constant-volume constraint. Hocking [13] shows an equivalence between a boundedness condition derived from the evolution equation for the interfacial shape of a two-dimensional drop. Lacey [16] poses a similar condition with his formulation more closely resembling (2.9).

In all these cases the formulation of the boundary condition is directly connected with the evolution equation. This equation is developed by solving for the velocity components (u, v, w) and substituting these into the kinematic boundary condition (2.4). The resulting partial differential equation governs the interfacial shape $h(x, z, t)$. With this in mind the equivalence between these conditions can be established.

To show this we integrate the continuity equation (2.3) across the thickness of the fluid system and use condition (2.5) to rewrite the kinematic condition (2.4) as

$$h_t + \nabla_1 \cdot \underline{Q} = 0, \tag{2.24}$$

where the horizontal volume flux $\underline{Q}$ is given in eqn. (2.8), and

$$\nabla_1 = \underline{i}\frac{\partial}{\partial x} + \underline{k}\frac{\partial}{\partial z}. \tag{2.25}$$

As shown in Fig. 3 consider a drop whose contact line is given in cylindrical coordinates by $r = A(\Psi, t)$. The volume of the drop V_c is given by

$$V_c = \int_0^{2\pi} \int_0^{A(\Psi,t)} hr\, dr d\Psi. \tag{2.26}$$

Differentiate eqn. (2.26) with respect to time using Leibnitz' rule and the condition of contact

$$h(A, \Psi, t) = 0 \tag{2.27}$$

to obtain

$$\int_0^{2\pi} \int_0^{A(\Psi,t)} h_t r\, dr d\Psi = 0. \tag{2.28}$$

This follows since the volume is a constant. Now integrate form (2.24) over the wetted area of the solid and apply the divergence theorem to find

$$\int_{A_{ws}}\int h_t d_A + \oint_{CL} [\underline{Q} \cdot \underline{\nu}]_{CL} ds = 0. \tag{2.29}$$

The first term of this expression is zero by condition (2.28) so that we must have

$$\oint_{CL} [\underline{Q} \cdot \underline{\nu}]_{CL} ds = 0. \tag{2.30}$$

This must hold for all positions of the contact line so the integrand $\underline{Q} \cdot \underline{\nu} = 0$ at the contact line. Thus, we recover (2.9). By evaluating $\underline{Q}$ and $\underline{\nu}$ from Hocking [13] and Lacey [16], one recovers their boundedness conditions at the contact line. Similarly, Greenspan [11] is able to abandon form (2.13) in lieu of form (2.26) since the two are equivalent to (2.9). We should mention that if one uses the zero-mass flux condition in form (2.9), when considering a drop configuration, mass is conserved. However, the initial volume of the drop must still be given in order to define the mass under consideration.

It is instructive to note the consequences of the zero-mass-flux condition as applied to an open (through flow) system such as the rivulet configuration. Let us integrate (2.24) over a fixed portion of a rivulet wetted area as shown in Fig. 4. We have

$$\int_{z_0}^{z_1} \int_{-B}^{A} h_t\, dx dz + \oint_S \underline{Q} \cdot \underline{\nu}\, ds = 0. \tag{2.31}$$

Now similar to the argument leading to eqn. (2.28) one can show that the first term of eqn. (2.31) is the time rate of change of the volume in this section of the rivulet

(2.32) $$\frac{dV}{dt} = \int_{z_0}^{z_1} \int_{-B}^{A} h_t \, dx dz.$$

We evaluate the line integral to find

$$\oint_S \underline{Q} \cdot \underline{n} \, ds = \int_{x=-B} \underline{Q} \cdot \underline{\nu}_B \, ds + \int_{-B}^{A} [\underline{Q} \cdot \underline{k}]_{z=z_1} \, dx + \int_{x=A} \underline{Q} \cdot \underline{\nu}_A \, ds$$

$$+ \int_{A}^{-B} [\underline{Q} \cdot (-\underline{k})]_{z=z_0} (-dx) = [\int_{-B}^{A} \int_{0}^{h} w \, dy dx]_{z=z_1}$$

(2.33) $$-[\int_{-B}^{A} \int_{0}^{h} w \, dy dx]_{z=z_0} = V_{\text{out}} - V_{\text{in}}$$

where we have used (2.9) to cancel the contributions at the contact lines. Therefore, eqn. (2.31) states that

(2.34) $$\frac{dV}{dt} = V_{\text{in}} - V_{\text{out}},$$

which is a mass balance. Thus, condition (2.9), whether for a closed or an open system, is consistent with conservation of mass, and in fact assures it. The set of boundary conditions (2.1), (2.2), and (2.9) defines a well-posed problem for flows possessing moving contact lines.

2d. Conclusions. The purpose of this section is to consider the type and number of boundary conditions necessary to describe a flow with moving contact lines. The most important aspect of the mathematical models developed here is the formalization of the contact-line boundary conditions. We find that the following three types of conditions are sufficient to form a well-posed problem:

i) Condition of contact
ii) Condition of contact angle as a function of contact-line speed
iii) Condition of zero-mass flux across the contact line.

The zero-mass flux condition is often left unposed explicitly. It is equivalent to other forms such as volume constraints and boundedness conditions, when the configurations are special. It is equivalent to a new kinematic condition on the contact line, viz.

(2.35) $$A_t + wA_z = u$$

that may be easier to implement when solving an evolution equation numerically. Whether one uses the form (2.35) or the form

(2.36) $$\underline{Q} \cdot \underline{\nu} = 0$$

or any of the volume constraint or boundedness forms already tried, the overall effect is to guarantee the condition that fluid does not leak out of the contact line.

3a. Asymptotic analyses for solidification systems. During the past ten years there has been a considerable effort in the modeling of systems involving a

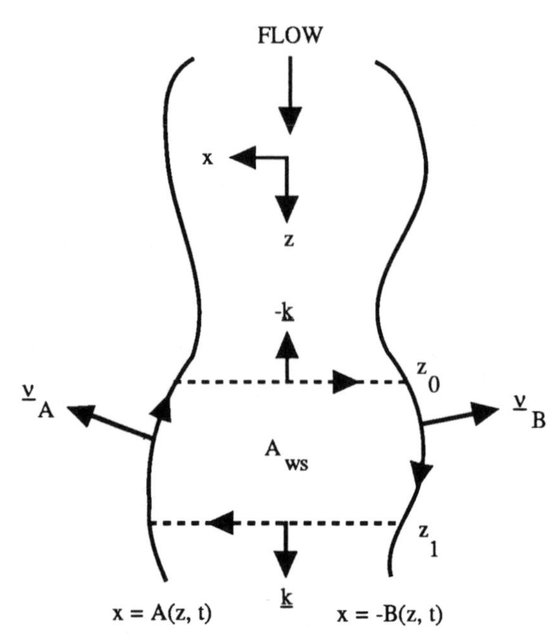

FIG. 4. *Sketch of a portion of the wetted area of a solid surface caused by a rivulet flow.*

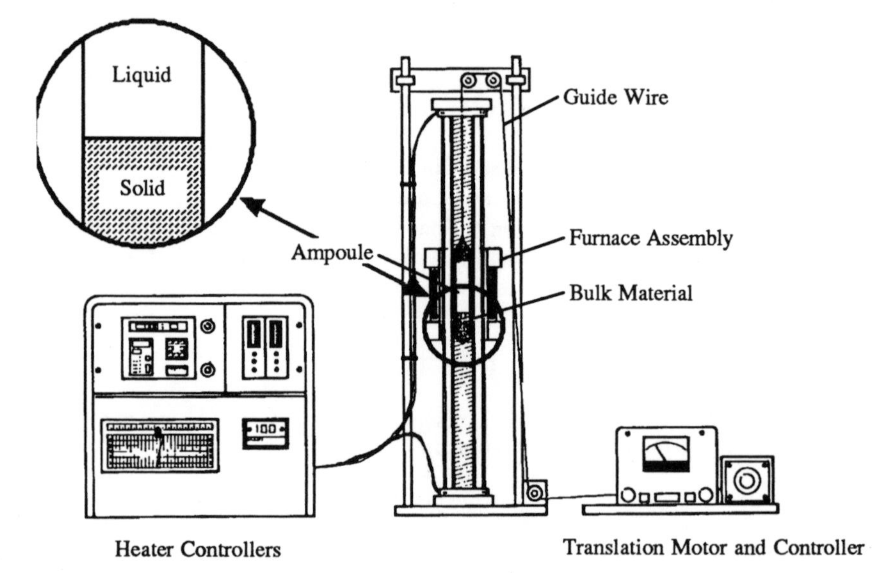

FIG. 5. *Sketch of a directional solidification apparatus.*

change of phase from the liquid to solid state. This section and the next will focus on two asymptotic approaches for investigating solidification systems. Figure 5 depicts an experimental arrangement used to grow crystals. In simplest terms an ampoule of melt is passed through a temperature gradient, hot to cold, resulting in the establishment of a solidification front. For sake of argument we assume that the melt is binary, composed of a solvent and a solute. Further we assume that not all of the solute is frozen into the solid, but rather some is rejected back into the liquid upon solidification. The goal of the crystal grower is to control the environment within which the solidification is taking place, so as to produce a crystal which is chemically homogeneous with respect to the solute, and as structurally perfect as possible. In these processes a variety of flows may develop. These will alter the solute and thermal fields, influencing the homogeneity and structure of the crystal. Such flows may be classified as:

1. Buoyancy driven – In this situation thermal and solutal gradients may alter the density of the melt. Hence, in the presence of gravity, flow results. Primarily radial thermal gradients are responsible for this type of flow, since the axial gradients are typically stabilizing. The rejection of solute lighter than solvent may lead to a double-diffusive convection.
2. Phase Change – If the density of the solid phase is larger than that in the liquid phase, the solid will shrink upon solidification. The melt then flows to fill the void. The opposite happens when the density of liquid is higher.
3. Forced Convection – In many situations, such as the Czochralski and float-zone configurations, a forced flow develops due to rotation of the growing crystals. Such rotation is intended to smooth out thermal asymmetries in the melt. Magnetic fields may also be applied to increase or dampen the flow in conducting melts.
4. Surface-Tension-Driven – The Czochralski and float-zone melts are confined by a liquid-gas interface. Variations in surface tension due to thermal and solutal gradients along the meniscus may lead to so-called Marangoni flows.

TABLE 1

Selected system, operating, and material parameters

Melt Height	$L = 10\text{cm}$
Velocity of Solidification	$V = 10^{-3} \frac{\text{cm}}{\text{s}}$
Thermal Diffusivity	$\kappa = 10^{-1} \frac{\text{cm}^2}{\text{s}}$
Solutal Diffusivity	$D = 10^{-5} \frac{\text{cm}^2}{\text{s}}$
Melt Viscosity	$\nu = 10^{-3} \frac{\text{cm}^2}{3} \text{s}$

To study the influence of flow on the solidification front, which determines the properties of the crystal, asymptotic approaches as well as numerical simulations focus on two types of models for the solidification system. These models are based on the fact that there are many different length scales inherent to the system. Table 1 lists several system, control, and material parameters in a typical semiconductor system.

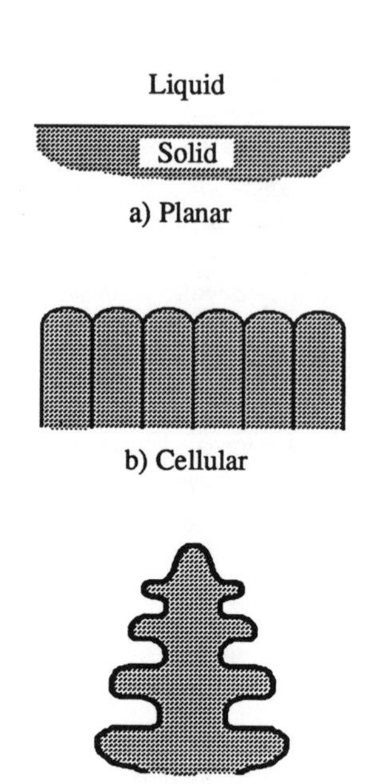

FIG. 6. *Sketch of interfacial shapes developing during solidification.*

From this list we identify the following four length scales:

(3.1)
$$\begin{aligned} &\delta_C = \tfrac{D}{V} = \text{solute diffusion boundary layer} \\ &\delta_v = \tfrac{\nu}{V} = \text{viscous length scale} \\ &L = \text{system geometry length scale} \\ &\delta_T = \tfrac{\kappa}{V} = \text{thermal diffusion length} \end{aligned}$$

These are ordered as

(3.2)
$$\delta_C \ll \delta_v \ll L \ll \delta_T$$

The first type of model is based upon the solute scale δ_C. The purpose of such models is to examing the "micro" scale morphology of the interface. It is well known (Mullins & Sekerka [19]) that a flat planar solidification front becomes unstable to a cellular structure as the solidification velocity is increased, keeping all other parameters fixed. Typical experimental observations are sketched in Figure 6. A further increase in velocity results in the cellular structure becoming dendritic. The mechanism for the latter is not understood. However, the cellular mechanism is well characterized by constitutional supercooling arguments. The rejected solute leads to a solute diffusion boundary layer ahead of the solidifying front. From thermodynamics the presence of

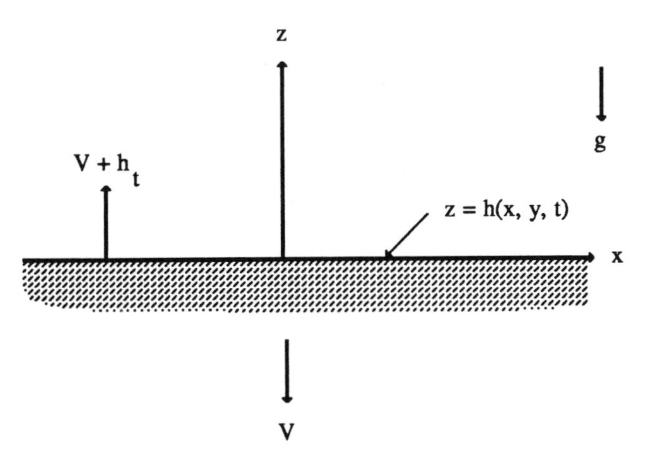

FIG. 7. *Sketch of an unbounded solidifying front. V is the velocity of solidification.*

this solute leads to an undercooling of the melt temperature below the equilibrium melting point, resulting in instability. Current investigations are examining the influence of melt flow on this morphological instability of the front. Whether convection sweeps solute so as to homogenize its distribution, resulting in a stabilizing effect, or sweeps it to reinforce the instability are current areas of research. Davis [6] reviews work in this area.

The second type of model is based upon the system length scale L. Such models intend to investigate the overall heat, mass, and momentum transport within the system and understand the influence of this transport on the macroscopic properties of the crystal.

We shall now present a brief review of asymptotic approaches towards solving both types of models.

3b. "Micro" scale models. Given that $\delta_C \ll L$, the system geometry may be taken as an unbounded domain, as shown in Figure 7. The governing equations are:

Liquid

$$\rho_0 \left(\frac{\partial \mathbf{v}}{\partial t} + \mathbf{v} \cdot \nabla \mathbf{v} \right) = -\nabla p + \mu \nabla^2 \mathbf{v} + \rho_0 \, g[1 - \beta_T (T - T_0) - \beta_c (c - c_0)]\mathbf{k} \tag{3.3}$$

$$\nabla \cdot \mathbf{v} = 0 \tag{3.4}$$

$$\frac{\partial T}{\partial t} + \mathbf{v} \cdot \nabla T = \kappa \nabla^2 T \tag{3.5}$$

$$\frac{\partial c}{\partial t} + \mathbf{v} \cdot \nabla c = D \nabla^2 c \tag{3.6}$$

Solid

$$\frac{\partial T_s}{\partial t} + V T_{sz} = \kappa_s \nabla^2 T_s. \tag{3.7}$$

Here ρ = density, $\mathbf{v}$ = fluid velocity, μ = viscosity, g = gravity, T = temperature in the liquid phase, β_T = thermal expansion coefficient, β_C = solute expansion coefficient, κ = thermal diffusivity, D = solute diffusivity, c = concentration of the solute, T_s = temperature in the solid phase, and V = velocity of solidification.

The Boussinesq approximation has been used for the state equation relating density variation with temperature and concentration, i.e.

$$\rho = \rho_0[1 - \beta_T(T - T_0) + \beta_C(c - c_0)]. \tag{3.8}$$

We also mention that typically the concentration equation is neglected in the solid phase since the solutal diffusivity in this phase is much smaller than the liquid phase. However, this cannot be done in situations involving melting or severe distortion of the solidifying front (Ungar & Brown [26]). Boundary conditions for the above are prescribed in the far field and at the solidifying front. We take

$$c \to c_\infty \text{ as } z \to \infty \tag{3.9}$$

$$T_z \to G \text{ as } z \to \infty \tag{3.10}$$

$$T_{sz} \to G_s \text{ as } z \to -\infty$$

in the far field. The far field concentration, c_∞, is taken to be uniform. G and G_S are prescribed temperature gradients. At the solidifying front we pose:

$$T = T_s \tag{3.12}$$

$$T = mc + T_M\left[1 + \frac{\gamma K}{\mathcal{L}}\right] \tag{3.13}$$

$$c_s = kc \tag{3.14}$$

$$c(1-k)(V + h_t)\underline{n} \cdot \underline{k} = -D\nabla c \cdot \underline{n} \tag{3.15}$$

$$\mathcal{L}(V + h_t)\underline{n} \cdot \underline{k} = [k_s \nabla T_s - k\nabla T] \cdot \underline{n} \tag{3.16}$$

$$(\rho_L - \rho_s)V(\underline{n} \cdot \underline{g}) = \rho_L(\underline{v} \cdot \underline{n}) \tag{3.17}$$

These represent continuity of temperature, thermodynamic equilibrium, segregation, mass balance, thermal balance, and no slip, respectively. A discussion of these equations can be found in Langer [17], for example. For purposes of this paper it is important to note the latent heat, $\mathcal{L}$, the segregation coefficient, k, and the surface free energy γ which multiplies the interfacial curvature K. The latter acts to stabilize the cellular structures discussed in Figure 6.

To illustrate an asymptotic approach to examining the above system we consider the Stefan problem in the domain of Figure 8. We introduce a characteristic length scale L and define the following scalings:

$$U = \frac{T - T_w}{T_M - T_w}, \quad \tau = \frac{k(T_M - T_w)}{L^2 \mathcal{L}} t, \quad \epsilon = \frac{k(T_M - T_w)}{\kappa \mathcal{L}} \ll 1 \tag{3.18}$$

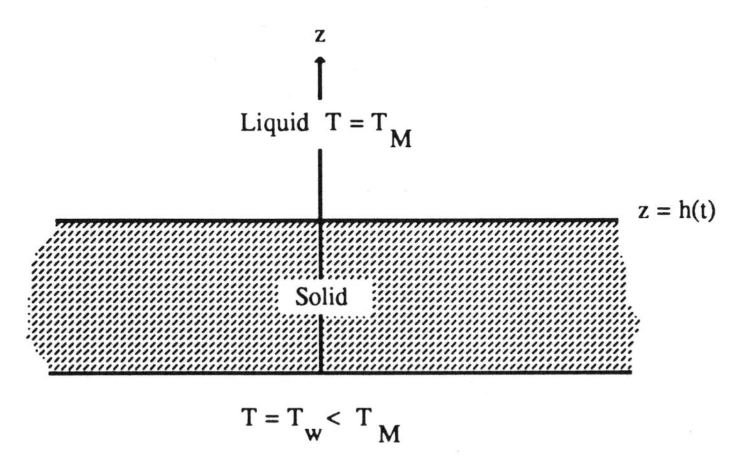

FIG. 8. *Sketch of the liquid/solid domain for the Stefan problem. The liquid phase temperature is fixed at the melting point T_M. The wall ($z = 0$) is fixed at a temperature $T_w < T_M$.*

The time scale is chosen to balance the release of latent heat, eq. (3.16), with the conduction of this heat in the solid phase. The Stefan problem assumes that the temperature in the liquid phase is maintained constant at the pure system melting point, T_M. The resulting scaled equations for the Stefan problem are

$$U_{zz} = \epsilon U_\tau \tag{3.19}$$

$$U(z = 0) = 0 \tag{3.20}$$

$$U(z = h) = 1 \tag{3.21}$$

$$h_\tau = U_z \text{ at } z = h \tag{3.22}$$

$$U(\tau = 0) = 1 \tag{3.23}$$

Here ϵ may be quite small due to the size of the latent heat, $\mathcal{L}$. There is a well known exact solution to the above given by

$$U = \frac{\operatorname{erf}\left(\frac{z\sqrt{\epsilon}}{2\sqrt{\tau}}\right)}{\operatorname{erf}(a)} \tag{3.24}$$

$$h = \frac{2a\sqrt{\tau}}{\sqrt{\epsilon}}, \text{ where } \operatorname{erf}(a) = \epsilon\frac{e^{-a^2}}{a\sqrt{\pi}} \text{ defines the constant } a. \tag{3.25}$$

Note that for $\epsilon \ll 1$, we have that $a \sim \sqrt{\frac{\epsilon}{2}}$ so that $h \sim \sqrt{2\tau}$. These follow from the asymptotic expansion of the error function for small argument.

Now consider the same problem but this time assume a straight-forward asymptotic expansion for the temperature U:

$$U = U_0(z,\tau) + \epsilon U_1 + \dots \tag{3.26}$$

Substituting this expression into the governing system, one finds the leading order problem and solution:

$$O(1): \begin{array}{l} U_{0zz} = 0 \\ U_0(z=0) = 0 \\ U_0(z=h) = 1 \end{array} \qquad \Rightarrow \qquad U_0 = \tfrac{z}{h} \tag{3.27}$$

Now substituting the expression for U_0 into the latent heat equation (3.22) we find the following isolated mathematical description, an evolution equation, for the free surface h

$$h_t - \frac{1}{h} = 0. \tag{3.28}$$

Solving this, using the initial condition, eq.(3.23), one recovers the asymptotic approximation of the exact answer. This model problem motivates the idea that an approach for studying the morphology of the solidifying front is to identify key limits within the full governing equations which allow the development of an evolution equation for the front. The above example illustrates the general idea by which one boundary condition is saved, in this case the latent heat balance, from which the evolution equation arises.

Several limits on the full governing system, yielding evolution equations, have been identified. Each of these limits involves a local analysis about a critical condition for the onset of instability in the planar base state. This state is characterized by no melt motion, sufficiently small latent heat and equal conductivities. It can be shown (Young and Davis [30]) that these assumptions lead to a linear temperature profile and an exponential concentration profile. Mullins and Sekerka [19] examined the linear stability of this base state and found the following expression for the growth rate σ:

$$\sigma = (1 - M^{-1} - a^2\Gamma)\left\{\left(\frac{1}{4} + a^2 + \sigma\right)^{\frac{1}{2}} + k - \frac{1}{2}\right\} - k \tag{3.29}$$

In the above, a is the disturbance wave number,

$$\Gamma = \frac{kT_M\gamma V}{D\mathcal{L}m(k-1)c_\infty} \tag{3.30}$$

$$M = \frac{m(k-1)c_\infty V}{kDG}. \tag{3.31}$$

Increasing Γ is stabilizing while $M^{-1} < 1$ leads to instability. M measures the undercooling due to the concentration gradient versus the applied thermal gradient G. Riley and Davis [21] sketched a plot of the marginal stability surface at the critical wave number. This is shown in Figure 9. Sivashinsky [24] was the first to

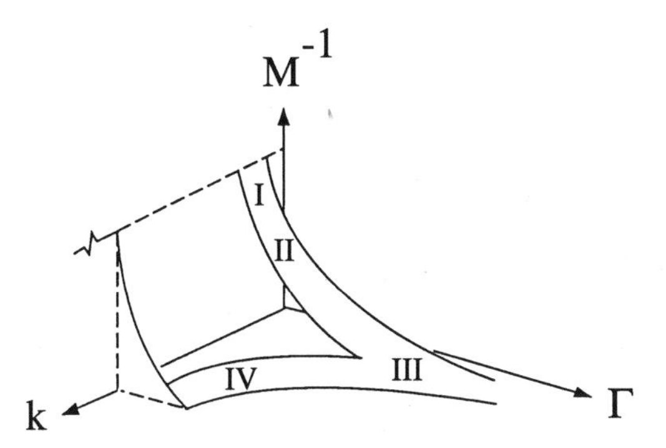

FIG. 9. *The marginal stability surface associated with transition from a planar to a cellular solidification front. Regimes subject to long-wavelength instabilities are*

I. $k \ll a^2 \ll 1,\ M^{-1} \sim 1,\ \Gamma \sim 1:$ Sivashinsky[24]
II. $k \sim a^2 \ll 1,\ M^{-1} \sim 1,\ \Gamma \gg 1:$ Riley and Davis[21]
III. $a^2 \ll k \ll 1,\ M^{-1} \ll 1,\ \Gamma \gg 1:$ Riley and Davis[21]
IV. $a^2 \ll k \sim 1,\ M^{-1} \ll 1,\ \Gamma \sim k^{-1}:$ Brattkus and Davis[2]

realize that there existed a regime on the $(M^{-1},\ \Gamma,\ k)$ surface which was subject to a long-wavelength (λ) instability

$$a = \frac{\frac{D}{V}}{\frac{\lambda}{2\pi}} \ll 1\,. \tag{3.32}$$

This is a mathematically attractive limit for investigation since the disparity in the vertical length scale ($\frac{D}{V}$ – diffusion boundary layer) versus the horizontal length scale (wavelength λ) leads to a reduction in the dimensionality of the problem. As shown in Figure 9 there are three other limiting cases which lead to long wavelength instabilities of the solidifying front. If one examines this limit, the following general comments apply:

a) The limits on the segregation coefficient, k, are physically relevant. Systems do exist for which k is small, $k = O(0.01)$.
b) The limits on M^{-1} are also relevant. M^{-1} near unity places one near the onset of instability. $M^{-1} \ll 1$ is in the absolute stability region. Here growth rates V are so large that the stabilizing effects of surface free energy overwhelm the destablizing effects of undercooling. As a result the cellular instability disappears.
c) The limits on the surface free energy are very restrictive. For most, if not all, materials, Γ is a very small number. As a result, it is not feasible to compare the quantitative predictions made by the evolution equations with

experiment. All experiments to date do not operate in a parameter regime consistent with this limit.

Thus the use of these types of evolution equations as a predictive tool is limited in a quantitative sense. However, qualitatively they predict many features which are consistent with experimental observation and numerical simulations of the full system. Hence, one is able to mathematically analyze regions I - IV and predict behavior of systems outside these regimes. As such the evolution equations have much potential for investigating new physics which can be included in the full system. Perhaps, this will be their primary role in future investigations.

Listed below is a brief summary of some of the evolution equations which have been derived:

Region I. This has been the most popular limit investigated. Such evolution equations, in general, are weakly nonlinear (small amplitudes for h) and predict subcritical bifurcation from the planar base state. No stable steady solutions exist to these equations without the inclusion of additional curvature terms (Kurtze [15] and Hyman et. al. [14]). Perhaps rational function approaches suggested by Hyman et. al. [14] may be developed to alleviate the difficulties in uniformity which develop when exponential solutions are expanded using Taylor series. Some evolution equations of this type are:
Sivashinsky [24]

$$h_t + \Gamma h_{xxxx} + [(1-h)h_x]_x + kh = 0 \tag{3.33}$$

Novick-Cohen and Sivashinsky [20]

$$h_t + \Gamma h_{xxxx} + [(1-h)h_x]_x + kh + \text{Latent Heat Correction} = 0 \tag{3.34}$$

Young and Davis [30]

$$h_t + \{\Gamma h_{xxxx} + [(1-h)h_x]_x\} \left[1 - \frac{R_{as}}{2(1+S_c^{-1})}\right] + kh = 0 \tag{3.35}$$

Young, Davis, and Brattkus [31]

$$h_t + \Gamma h_{xxxx} + \mu h_{xxx} + [(1-h)h_x]_x + kh = 0 \tag{3.36}$$

Hyman, Novick-Cohen, and Rosenau [14], Kurtze [15]

$$h_t + [\Gamma h_{xx}(1+\epsilon^3 h_x^2)]_{xx} + [(1-h)h_x]_x + kh = 0 \tag{3.37}$$

Young and Davis [32]

$$h_t + \Gamma h_{xxxx} + [(M_c^{-1} - h)h_x]_x + kh = 0 \tag{3.38}$$

Regions II - IV. These limits yield evolution equations with more attractive features than the above limit. Here strongly nonlinear equations (h of arbitrary amplitude) predicting supercritical bifurcation and three-dimensional hexagonal node structures are possible. Such predictions are consistent with the numerical simulations of McFadden, et. al. [18].

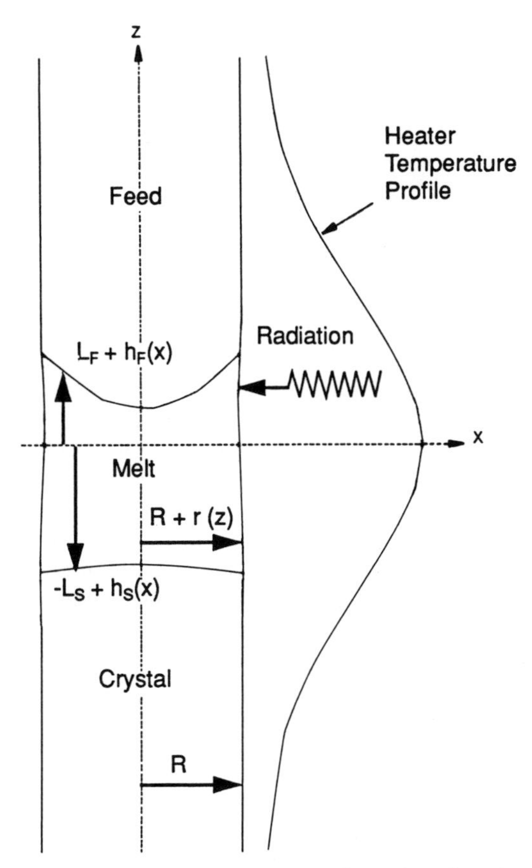

FIG. 10. *Sketch of the float-zone crystal growth configuration.*

Equations of this type have also predicted long-wave interactions in morphological and convective instabilities (Riley and Davis [22]). Evolution equations of this type are:
Brattkus and Davis [2]

$$\text{(3.39)}\qquad \begin{aligned} &h_{tt} - \nabla^2 h_t + \tfrac{1}{4}(1-\nu^2)\nabla^4 h + \nabla^2 h + kh = h_t \nabla^2 h + |\nabla^2 h|_t^2 \\ &-\tfrac{1}{2}(1-\nu)\nabla^2(|\nabla h|^2) - \nu\nabla\cdot(\nabla^2 h \nabla h) - \tfrac{1}{2}\nabla\cdot(|\nabla h|^2 \nabla h) \end{aligned}$$

Riley and Davis [22]
(3.40)

$$h_t - M\nabla^2 h_t + M\nabla^4 h + [M - 1 - kM]\nabla^2 h + kh = \nabla\cdot(h\nabla h) - M\nabla\cdot(\nabla^2 h \nabla h)$$

4. Float-zone example–"macro" scale model. This last example is a model of a very complicated solidification system, the float-zone technique. The details of the model will not be presented. They can be found in Young and Chait [33]. Here we intend to illustrate the use of domain perturbations to construct a model. In this float-zone (FZ) model we note that there are twenty-three independent material, system, and operating parameters. While numerical simulations may be able to investigate a wide range of magnitudes of these parameters, in order to achieve an asymptotic solution very subtle balances amongst the parameter orderings must be made to achieve a reasonable solution. We shall highlight one such balance to illustrate the mathematical and physical issues which come to play in the ordering decision.

Figure 10 depicts the float zone configuration. A heater establishes a molten zone bounded by a free melting interface, a free solidifying interface, and a free liquid/gas meniscus. Further the width of the growing crystal is also free. This width is controlled by a contact angle mechanism as discussed in Surek and Coriell [25]. Modeling objectives are to examine the coupled dynamics of heat, melt, and solute transport with consistent determination of the melting, solidifying, and liquid/gas free surfaces. We shall focus on development of an isolated mathematical description of the mean planar positions L_F and L_S. Knowledge of the location of these interfaces is important since too much heat in the zone will lead to a large distance $L = L_S + L_F$ potentially resulting in a loss of the meniscus due to a capillary instability, while too little heat results in no melt through.

The solution procedure of Young and Chait [33] involves a domain perturbation, schematically shown in Figure 11. This thin domain approximation allows one to reduce the dimensionality of the model, similar to the long-wavelength approach. However, it will lead to boundary layer analyses as shown in Figure 12. Such procedures have been used previously for analysis of similar systems. Sen and Davis [23] examine Marangoni convection in a slot, while Brattkus and Davis [1] investigate buoyancy driven flow in a Bridgman type configuration as shown in Figure 5. These thin domain analyses are very attractive mathematically and are quite relevant since many systems exhibit an underlying geometry which might be considered as thin.

We shall not list the entire set of governing equations for the FZ model. Instead we list the small subset of equations for heat transfer within the melt zone. These are:

$$\text{(4.1)}\qquad \epsilon^2 M_a[-uT_x + wT_z] - \epsilon^2 P_e T_z = T_{xx} + \epsilon^2 T_{zz}\,.$$

At the liquid gas interface

$$\frac{-[T_x - \epsilon^2 T_z r_z]}{\sqrt{1+\epsilon^2 r_z^2}} = R_L[T^4 - \Theta^4] + B_i[T - \Theta] \tag{4.2}$$

where Θ = the temperature profile of the heater.
At the Centerline

$$T_x = 0\,. \tag{4.3}$$

At the solidification front

$$T = Mc + T_M^* \tag{4.4a}$$

$$\epsilon P_e S_t = K[\epsilon T_{Sz} - T_{Sx} h_{Sx}] - [\epsilon T_z - T_x h_{Sx}]\,. \tag{4.4b}$$

If one tries a straight forward expansion, $T = T_0 + \epsilon T + \ldots$, we find at leading order

$$T_{0xx} = 0 \tag{4.5}$$

$$-T_{0x} = R_L[T^4 - \Theta^4] + B_i[T - \Theta] \text{ at } x = 1 \tag{4.6}$$

$$-T_{0x} = 0 \text{ at } x = 0 \tag{4.7}$$

$$T_0 = T_M \text{ at } z = -L_S \tag{4.8}$$

$$\epsilon P_e S_t = K[\epsilon T_z - T_{Sx} h_{Sx}] - [\epsilon T_z - T_x h_{Sx}] \text{ at } z = -L_S\,. \tag{4.9}$$

From equations (4.5) and (4.7) we note that $T_0 = T_L(z)$ a function of z only. Substituting into equation (4.6), the surface heat transfer condition, we find that

$$0 = R_L[T_L^4 - \Theta^4] + B_i[T_L - \Theta] \tag{4.10}$$

The only solution is $T_L = \Theta$. In other words the temperature of the melt is identical to the temperature of the heater. A perfectly fine mathematical solution, but one which is not physically relevant. The heat transfer parameters R_L and B_i are so large that too much heat transfers into the melt. Hence we must order their magnitudes to be smaller. Should they be $O(\epsilon)$, $O(\epsilon^2)$, or $O(\epsilon^{\frac{1}{2}})$, ... quantities? We select them to be $O(\epsilon^2)$. This prescription is not ad hoc but based upon the fact that the most interesting situation would be when convective, conductive, and surface heat transport all compete. From equation (4.1) we see that convective and conductive transport are $O(\epsilon^2)$ contributions. Thus we order R_L and B_i to be $O(\epsilon^2)$ as well. Similar arguments are used throughout the development of the FZ model.

As a result of the $O(\epsilon^2)$ orderings one can solve for the $O(\epsilon^2)$ contribution to the temperature field. In order to satisfy the heat balance, equation (4.6) the following ODE must be solved for the unknown $T_L(z)$:
Melt ($-L_S < z < L_F$)

$$(1 + r_0)(T_{Lzz} + P_e T_{Lz}) + T_{Lz} r_{0z} - \overline{B}_L T_L - \overline{R}_L T_L^4 = -\overline{B}_L \Theta - \overline{R}_L \Theta^4 \tag{4.11}$$

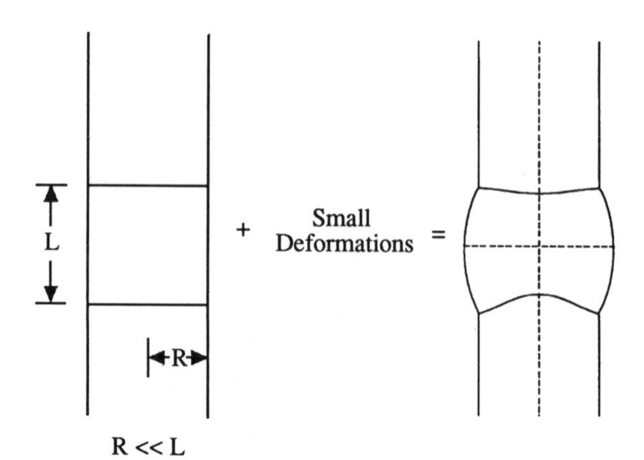

FIG. 11. *Sketch of the domain perturbation for the float-zone model.*

$$T_L(-L_S) = T_L(L_F) = T_M^* \tag{4.12}$$

Feed ($z > L_F$)

$$T_{F0zz} + P_S T_{F0z} - \overline{B}_S T_{F0} - \overline{R}_S T_{F0}^4 = -\overline{B}_S \Theta - \overline{R}_S \Theta^4 \tag{4.13}$$

$$T_{F0}(L_F) = T_M^*, \; T_{F0}(L_2) = T_a \tag{4.14}$$

Solidifying Crystal ($-L_S > z$)

$$T_{S0zz} + P_S T_{S0z} - \overline{B}_S T_{S0} - \overline{R}_S T_{S0}^4 = -\overline{B}_S \Theta - \overline{R}_S \Theta^4 \tag{4.15}$$

$$T_{S0}(-L_S) = T_M^*, \; T_{S0}(L_1) = T_a \tag{4.16}$$

The other equations shown above result from similar analyses in the feed and growing crystal. Augmenting this system with the heat balances from equation (4.9)

$$K T_{S0z}(-L_S) - T_{Lz}(-L_S) = P_e S_t \tag{4.17}$$

$$K T_{F0z}(L_F) - T_{Lz}(L_F) = P_e S_t \tag{4.18}$$

yields an isolated mathematical description for the free surface planar positions L_S and L_F. This is still a free boundary problem but much simpler than the full system. It is solved numerically because of the inclusion of the radiation terms containing temperature raised to the fourth power. However this numerical simulation takes less than a fraction of a second of CPU time.

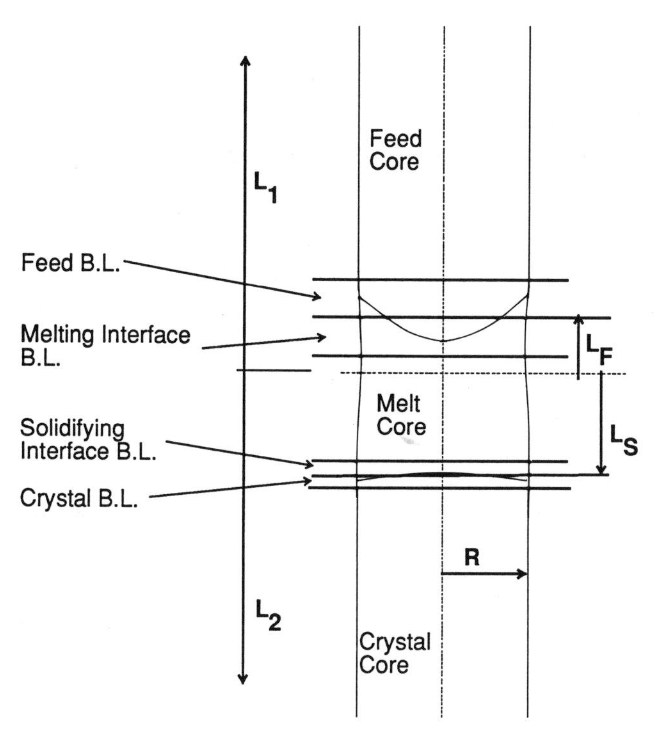

FIG. 12. *Sketch of the boundary layers resulting from the limit* $R \ll L = L_F + L_S$.

Proceeding from here one can calculate corrections to the planar front. The details are in Young and Chait [33]. Figure 13 depicts typical zones calculated from the asymptotic model. Recently Zhang [34] has extended the above analysis to an axisymmetric geometry. This model was then compared to the conduction dominated numerical simulation of Duranceau and Brown [7]. Qualitative agreement is excellent. Table 2 lists a quantitative comparison at the four locations shown in Figure 13. Again there is excellent agreement at the solidification front. More error occurs at the melting front. Although considering that $\epsilon = 1.2$ in the comparison, while the model was developed for $\epsilon \ll 1$, the quantitative agreement is remarkable. While this comparison is by no means an exhaustive study throughout parameter space, it does illustrate that at times one can push the asymptotic limits to extreme cases, yet achieve reasonable results. Such results in the above model are an indication that the correction terms are indeed quite small. Brown [3] has postulated that such is the case because the data for silicon, which was used for the comparison, indicates a low latent heat. Hence the simplified model is able to account for its transport. Accuracy would be lost for higher latent heat. This explanation is consistent with a degrading of accuracy at the melting front. Here the heater is closer to the front so melting occurs at a higher rate. Thus more latent heat is liberated to the solid and accuracy is diminished as a result.

Nevertheless, this model demonstrates the potential of asymptotic simulations. Perhaps one might speculate that problems which are currently not computationally feasible or prohibitively expensive, may be treated by approaches incorporating asymptotic solutions in certain regions of the domain, with numerical simulation in the others.

5. Summary. We have examined three models of systems which involve viscous free surface flows. The first model concerned the development of boundary conditions at a contact line. Conditions of contact, contact angle as a function of contact-line speed, and zero-mass flux across the contact line were posed. The description of the latter condition unified several different approaches to the formulation of contact-line boundary conditions.

The second and third models are solidification systems. Both models involved asymptotic solution methodologies designed to isolate a mathematical description of the solidifying front. One concerned the development of an evolution equation for the solidification front, valid on a micro scale. The other, valid on a macro scale, concerned the development of a system of equations to locate the planar positions of melting and solidifying fronts, in a float zone. Both models illustrate that asymptotic approaches to the investigation of viscous free surface flows take advantage of disparity in length and time scales allowing one to reduce the complexity of the model, yet, potentially, retain much of the physics of the system. Such model reduction may lead to quantitative limitations. However, these do not necessarily preclude accurate qualitative predictions of instability and coupling behavior. This predictive capability gives asymptotics the role of a complement to computational simulations but not a substitute.

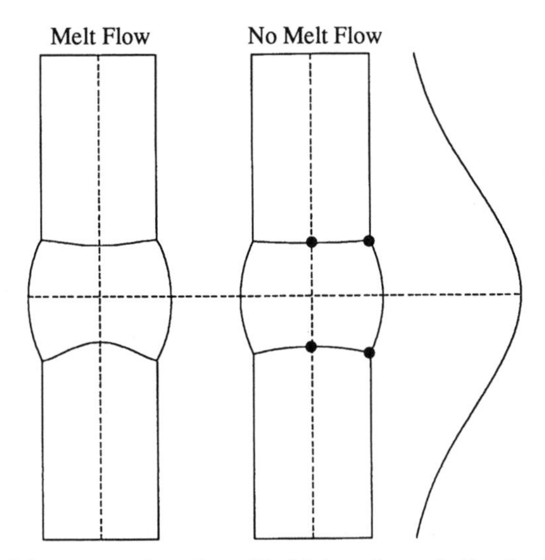

FIG. 13. *Calculated float zone configurations. The labels on the conduction dominated zone (no melt flow) correspond to the data of TABLE 2.*

TABLE 2

Comparison of conduction-dominated models

Interface	Duranceau and Brown [7]	Zhang [34]	% Error
$h_S(0)$	− 0.5352cm	− 0.5195	2.9
$h_S(1)$	− 0.6918	− 0.7448	7.7
$h_F(0)$	0.0915	0.1954	113
$h_F(1)$	0.4507	0.5200	12.5

REFERENCES

1. K. BRATTKUS AND S. H. DAVIS, *Directional solidification with heat losses*, J. Crystal Growth, 91 (1988), pp. 538-556.
2. K. BRATTKUS AND S. H. DAVIS, *Cellular growth near absolute stability*, Physical Review B, 38 (1988), pp. 11452-11460.
3. ROBERT A. BROWN, private communication.
4. S. H. DAVIS, *Moving contact lines and rivulet instabilities. Part 1. The static rivulet*, J. Fluid Mech., 98 (1980), pp. 225-240.
5. S. H. DAVIS, *Contact-line problems in fluid mechanics*, J. Appl. Mech., 50 (1983), pp. 977-982.
6. S. H. DAVIS, *Hydrodynamic interaction in directional solidification*, J. Fluid Mech., 212 (1990), pp. 241-262.
7. JACQUES L. DURANCEAU AND ROBERT A. BROWN, *Thermal-capillary analysis of small-scale floating zones*, J. Crystal Growth, 75 (1986), pp. 367-389.
8. E. B. DUSSAN V. AND S. H. DAVIS, *On the motion of a fluid-fluid interface along a solid surface*, J. Fluid Mech., 65 pp. 71-95.
9. E. B. DUSSAN V., *On the spreading of liquids on solid surfaces: static and dynamic contact lines*. Ann. Rev. Fluid Mech., 11 (1979), pp. 371-400.
10. E. B. DUSSAN V. AND R. T. -P. CHOW, *On the ability of drops or bubbles to stick to non-horizontal surfaces of solids*, J. Fluid Mech., 137 (1983), pp. 1-30.
11. H. P. GREENSPAN, *On the motion of a small viscous droplet that wets a surface*, J. Fluid Mech., 84 (1978), pp. 125-144.
12. H. P. GREENSPAN AND R. M. McCAY, *On the wetting of a surface by a very viscous fluid*, Stud. Appl. Math., 64 (1981), pp. 95-112.
13. L. M. HOCKING, *Sliding and spreading of thin two-dimensional drops*, Q. J. Mech. Appl. Math., 34 (1981), pp. 37-55.
14. J. M. HYMAN, A. NOVICK-COHEN, AND P. ROSENAU, *Modified asymptotic approach to modeling a dilute-binary-alloy solidification front*, phys. Rev. B, 37 (1988), pp. 7603-7608.
15. D. A. KURTZE, *Formation of deep grooves in directional solidification*, Phys. Rev. B, 37 (1988), pp. 370-378.
16. A. A. LACEY, *The motion with slip of a thin viscous droplet over a solid surface*, Stud. Appl. Math., 6D (1982), pp. 217-230.
17. J. S. LANGER, *Instabilities and pattern formation in crystal growth*, Rev. Modern Phys., 52 (1980), pp. 1-28.
18. G. B. McFADDEN, R. F. BOISVERT, AND S. R. CORIELL, *Nonplanar interface morphologies during unidirectional solidification of a binary alloy. II. Three-dimensional computations*, J. Crystal Growth, 84 (1987), pp. 371-388.
19. W. W. MULLINS AND R. F. SEKERKA, *Stability of a planar interface during directional solidification of a dilute binary alloy*, J. Appl. Phys., 35(1964), pp. 444-451.
20. A. NOVICK-COHEN AND G. I. SIVASHINSKY, *On the solidification front of a dilute binary alloy: thermal diffusivity effects and breathing solutions*, Physica D, 20 (1986), pp. 237-258.
21. D. S. RILEY AND S. H. DAVIS, *Long-wave morphological instabilities in the directional solidification of a dilute binary mixture*, SIAM J. Appl. Math., 50 (1990), pp. 420-436.
22. D. S. RILEY AND S. H. DAVIS, *Long-wave interactions in morphological and convective instabilities*, Applied Mathematics Technical Report No. 8829, Northwestern University.
23. A. K. SEN AND S. H. DAVIS, *Steady thermocapillary flow in two-dimensional slots*, J. Fluid Mech., 121 (1982), pp. 163-186.
24. G. I. SIVASHINSKY, *On the cellular instability of a dilute binary alloy*, Physica D, 8 (1983), pp. 243-248.
25. T. SUREK AND S. R. CORIELL, *Shape stability in float zoning of silicon crystals*, J. Crystal Growth, 37 (1977), pp. 253-271.
26. L. H. UNGAR, M. J. BENNETT, AND R. BROWN, *Cellular interface morphologies in directional solidification. III. The effects of heat transfer and solid diffusivity*, Phys. Rev. B, 31 (1985), pp. 5923-5930.
27. R. H. WEILAND AND S. H. DAVIS, *Moving contact lines and rivulet instabilities Part 2. Long waves on flat rivulets*, J. Fluid Mech., 107(1981), pp. 261-280.

28. G. W. YOUNG, *Dynamics and stability of flows with moving contact lines*, Ph.D. Thesis, Northwestern University, Evanston, Illinois, (1984).
29. G. W. YOUNG AND S. H. DAVIS, *A note on contact-line boundary conditions*, Applied Mathematics Technical Report No. 8411, Northwestern University, Evanston, Illinois, (1985).
30. G. W. YOUNG AND S. H. DAVIS, *Directional solidification with buoyancy in systems with small segregation coefficient*, Phys. Rev. B, 34 (1986), pp. 3388-3396.
31. G. W. YOUNG, S. H. DAVIS, AND K. BRATTKUS, *Anisotropic interface kinetics and tilted cells in unidirectional solidification*, J. Crystal Growth,83 (1987), pp. 560-571.
32. G. W. YOUNG AND S. H. DAVIS, *Morphological instabilities in directional solidification of a binary alloy: end effects*, SIAM J. Appl. Math., 49 (1989), pp. 152-164.
33. G. W. YOUNG AND A. CHAIT, *Surface tension driven heat, mass, and momentum transport in a two-dimensional float zone*, J. Crystal Growth, 106 (1990) pp. 445-466.
34. YAJUN ZHANG, *An asymptotic simulation of surface tension driven heat, mass, and momentum transport in an axially symmetric float zone*, M.S. Thesis, The University of Akron, Akron, Ohio, (1991).

THEORETICAL ISSUES ARISING IN THE MODELING OF VISCOUS FREE-SURFACE FLOWS

W.G. PRITCHARD*, PATRICIA SAAVEDRA†,
L. RIDGWAY SCOTT‡, and S.J. TAVENER*

Abstract. This paper discusses theoretical and computational issues regarding viscous flows which have a free surface. A number of mathematical models for a particular flow are described and compared, both with one another and with some physical experiments. We consider some approximate models based both on lubrication theory and finite element methods. The importance of the choice of boundary conditions in modeling practical flow phenomena is discussed, and some related open theoretical questions regarding the well-posedness of mathematical models for such phenomena are presented. The discussion also touches upon the role that surface tension has so far played in the mathematical theory of free-surface flows and in many numerical calculations. Briefly outlined is some preliminary work related to convergence estimates for finite-element methods for free-boundary problems.

Key words. Free boundary, fluid flow, finite-element method, lubrication approximation.

AMS(MOS) subject classifications. 76D05, 35Q10, 65N30

1. Introduction. Free-surface flows of viscous materials display a tremendous variety of interesting phenomena. Although inertial effects are relatively unimportant (or even negligible) for many of these flows, the presence of the free surface allows for the possibility of nonlinear effects and, accordingly, such flows can exhibit bifurcation phenomena, hysteresis effects, and nonlinear coherent structures. The theory and application of approximate models for free-surface problems for viscous fluid flows are developing rapidly. The subject has matured significantly in recent years, but there are still important issues that need to be addressed, especially with respect to the underlying theory.

In order to illustrate some of the achievements and the limitations of the current theoretical and computational models, we present here a survey of certain aspects of our recent researches, together with a discussion of several unresolved theoretical difficulties in the area. Many of the unresolved problems concern the well posedness of the underlying boundary-value problem, whereas others relate to the theoretical foundations of the approximate models.

One of the best known examples of the unexpected properties displayed by viscous free-surface flows is the so-called die swell phenomenon, a nice example of which is shown in the textbook by Lodge [20, p242], where the diameter of the jet emerging from a circular tube may expand to well over twice the diameter of the tube whence the jet emanated. While many of the interesting demonstrations of die swell are associated with the flow of non-Newtonian fluids, the corresponding Newtonian flow problem is of considerable interest in its own right, in that it provides important clues

* Department of Mathematics, Penn State University.

† Dept. de Matematicas, Universidad Autonoma Metropolitana, Mexico.

‡ Department of Mathematics, The University of Houston.

about how to pose mathematically the more general flow problem. Thus, while the ultimate motivation for some of the flows discussed here is that of understanding the mechanics of non-Newtonian materials, we shall henceforth concentrate on the (more explicitly defined) Newtonian problem.

Experimentally the Newtonian flow problem has an interesting history. For simplicity, we shall assume the diameter of the die to be unity. The early work of Middleman & Gavis [22] and Goren & Wronski [12] indicated that the diameter of the emerging jet increased from a value of $\sqrt{3}/2$ (as would be expected for a flow at very large Reynolds numbers) to a value of nearly 1 as the Reynolds number of the flow was decreased. Unfortunately the experiments could not be extended below a Reynolds number of approximately 4 because the liquid jet developed a considerable droop under the influence of gravity.

In some subsequent experiments by Batchelor, Berry & Horsfall [3] the liquid was extruded into a neutrally buoyant medium, thereby avoiding the earlier difficulties, and it was found that the jet *expanded* to approximately 1.13 times the diameter of the tube. The Reynolds numbers for these experiments were around 10^{-8} and the viscosity of the extrudate was so large that the surface-tension parameter $S := T/\mu U$ (where T is the surface tension, μ is the dynamical viscosity, and U is a characteristic velocity) was small. Attempts by Pritchard [28] to carry out the same kind of experiment produced, however, quite different results in that a uniform jet was not observed and instead a large, roughly spherical blob (resembling a balloon being inflated) developed at the end of the pipe from which the liquid was extruded (see figure 2 of the cited reference). The Reynolds numbers for Pritchard's [28] experiments were small (approximately 10^{-2}) but an important difference between [3] and [28] was that the surface tension parameter was of the order of 100 for Pritchard's experiment. A simple explanation for the different characters of these two experiments is that, in the latter experiment, the liquid was extruded at such a slow rate that the forces of surface tension dominated the viscous stresses, with the result that the extrudate adopted the observed, nearly spherical shape. The dramatic difference between these two experiments suggests that a mathematical model for this class of flows could involve some subtleties.

The experiments just described lead immediately to suggestions regarding the critical ingredients of well-posed mathematical models. The strong dependence on the shape of the extrudate from experiment to experiment implies that the boundary conditions can play an influential role in the outcome, a point we shall consider in more detail in §6. Thus, the modeling of manufacturing processes such as the blowing of "Saran" film, the spinning of man-made fibers and the continuous casting of aluminum requires the careful application of boundary conditions on the flow domain if the appropriate processes are to be represented faithfully by the mathematical model.

One very important application of viscous free-surface flows with significant similarities to the die-swell problem arises in the coating industry (cf. papers by Scriven *et al.* and Kistler in this volume), one particularly familiar example being the coating of photographic emulsion onto its backing strip. The emulsion must be transferred onto the backing strip in a layer of thickness of the order of 100Å, the uniformity of which is very important and the stratification of which is crucial so that, when used for photographic purposes, the various color components are filtered in the correct sequence and degree as light passes through the emulsion. A further application of this class of motions arises in the field of geophysics where, for example, gravity currents and laval flows have been modeled (e.g. see [14]) using a lubrication approximation.

A familiar property of viscous free-surface flows is the formation of singular points or lines in the free surface. It is quite common to see a bubble develop a conical point when, for example, mixing a jar of honey. But it is also appears experimentally to be possible to have two surfaces come together along a line and form a cusp-like singularity. The theoretical possibility of the formation of such cusps was considered in the early 1970's by Richardson [36, 37] and, more recently, Joseph *et al.* [17] have re-examined these issues and have provided some excellent photographic evidence for the formation of cusps. Singularities in the free-surface curvature arise commonly in free-surface flows (they are often a precursor to the formation of new bubbles) and as such constitute an important class of flow problems. Although a mathematical theory that could incorporate such phenomena is well beyond the current state-of-the-art, some of the ideas discussed subsequently could potentially provide key parts of the appropriate theoretical framework.

In the present paper we have limited our scope to steady free-surface flows but, because of their enormous practical and theoretical importance to the subject, it is worth making brief reference to some of the time-dependent flows that have been observed. Such flows have been observed with coherent (i.e. soliton-like) structures, and with time-periodic and even apparently chaotic motions. Many examples of time-dependent flows are given in Pritchard [29] and examples of flows with coherent structures have been observed in experiments relating to the so-called printer's instability (e.g. see [33]). Many of these time-dependent motions arise as bifurcations unfolded by tiny imperfections, such as a small misalignment of a surface or an axis from the horizontal, or of small nonuniformities between nearly parallel surfaces.

While the modeling processes and the mathematical formulation and properties of a given model are of crucial importance in providing a well-posed problem, it is also essential to be able to predict flow fields effectively. In the work to be described below we shall consider two such approaches to providing a good representation of the flow field. One of these, the familiar lubrication approximation, has proved to be quite efficient in predicting the shape of the free surface under certain conditions. The predictions from such a model can be used on their own, or as a starting guess for a direct numerical technique, such as a finite-element or a finite-difference method. Asymptotic models have a natural physical parameter that relates to the quality of the model, being exact only when the parameter tends to some limit. In practice, it is useful to understand the range of applicability of such models via comparisons with experiments (or other means) to quantify the dependence of model quality on the parameter.

Direct numerical models, such as those provided by finite-element methods, formally are only approximate models to the flow equations. However, the models involve an accuracy parameter, typically the mesh size, which can be varied to achieve, in principle, an arbitrary degree of quality. Sometimes the achievement of a high degree of accuracy (say of the order of 0.1%) can be quite demanding on computational resources, and it is useful to have some guide as to the size of gridding required in model problems. We have studied both of the above approaches and have compared them with laboratory experiments as well as with each other. We begin with a brief summary of the findings of those quantitative studies.

2. Benchmark problems. In order to make quantitative assessments of mathematical models, it is essential to have physical problems for which there is either an exact solution or experimental data. For example, the familiar "driven cavity" problem, which is frequently used as a benchmark problem for numerical schemes in domains having a known boundary, neither has an exact analytical solution nor is it experimentally realizable. In addition, it is important that test problems do not involve extraneous features (such as the introduction of spurious singularities through a truncation of an infinite domain) that would make an assessment of the models difficult. Finally, it is important that the problem be nontrivial in all important aspects. The latter requirement makes all known analytical solutions for free-boundary problems for viscous fluids of limited value. However, they can provide useful preliminary tests, and for completeness we list the ones of which we are aware.

The simplest free-surface problem is flow down a flat plate, having a parabolic profile for the flow field and a flat free-surface. A rigid body rotation yields a solution to a free-boundary problem with a circular (or cylindrical, in three dimensions) free surface and thus provides a useful test of the free-surface part of a solver. However, the velocity field is a linear function of the spatial variables, and this makes it essentially trivial for most finite-element methods, which would tend to have the flow field represented exactly. The static meniscus problem has a more complicated free-surface representation, but an even more trivial flow field (i.e., zero). Nevertheless, it is useful for testing attachment conditions at boundaries. All of these flows can be quite effective in assessing codes and finding bugs [30]. An additional problem of interest is related to Jeffery-Hamel flow [31] in which the flow profile is quite complex but the free-boundary is flat.

A simply defined physical problem is that of determining the flow in a film of fluid on the inside (or outside) of a rotating cylinder. For laboratory experiments, the length of the cylinder would of course have to be finite, but for a sufficiently long cylinder, some basic flows should be nearly two dimensional away from the end walls. However, for the more convenient (and interesting) case of flow on the inside of the cylinder, it is relatively difficult to measure the shape of the free surface [7]. Moreover, it has proved difficult with some codes we have tested to model this flow problem, one difficulty being that the free-boundary must be described by a periodic function instead of having known endpoint conditions.

An even simpler test problem is afforded by an experiment based on flow down a perturbed flat plate. Figure 1 shows the basic flow domain for a particular experiment, involving flow down a uniform channel perturbed by two neighboring bumps in the bed, at a Reynolds number of 36.6. (For the two-dimensional flows considered here, the Reynolds number is given by the ratio of the volume flux per unit width of the flow to the kinematical viscosity of the fluid.) The experiment depicted here (and described in more detail in [30]) has been chosen carefully for its simplicity, allowing accurate quantitative experiments to be carried out and also allowing an extremely close representation by an appropriate mathematical model through the modeling of

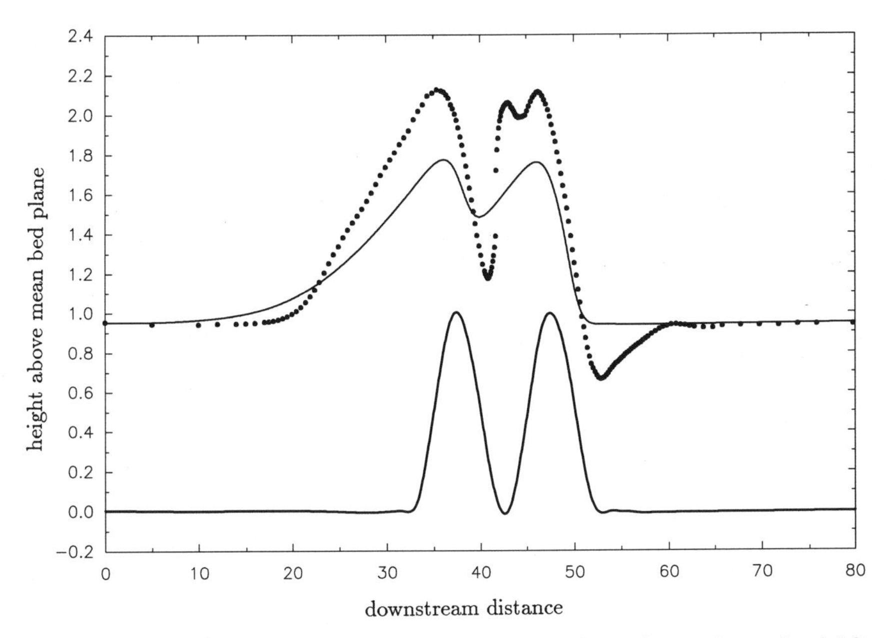

FIG. 1. *Comparison between experimental measurements (indicated by •) of the free-surface height above the plane defined by the uniform part of the channel (the channel bed is denoted by the lower solid curve) and the prediction based on a linear lubrication model (denoted by the upper solid curve). The experimental conditions had Re=36.6, S=1.66 and an upstream depth of 0.948.*

both the boundary conditions and the dynamical equations. The resolution of the flow problem and the comparisons with the empirical data have thus been chosen to be exemplary of a class of fluid motions, rather than to expose new physical phenomena.

A key feature of the flow down a perturbed flat plate is that it returns, exponentially rapidly, to the asymptotic state (given by the unperturbed case) away from the perturbations, as has been recently proved by Abergel & Bona [1]. This means that the inflow and outflow conditions are known to arbitrary accuracy sufficiently far upstream and downstream, and these can then be used as boundary conditions for a model on a bounded domain.

For the experiments shown herein, the uniform part of the channel bed was inclined at an angle $\alpha = 0.0735$ rad to the horizontal. Let x denote the coordinate direction parallel to the flat portion of the channel bed and let y denote the coordinate direction orthogonal to the flat portion of the bed. Both x and y are scaled by the maximum height (10.0 mm) of the bumps in the channel bed so that, for the results shown in the figures, distances in the downstream direction appear to be considerably foreshortened compared with those in the y direction. Comparisons of various models using these experiments will be described in the next two sections.

3. Lubrication models. Asymptotic methods can be quite effective in reducing the dimensionality of free-surface flow problems. For steady two-dimensional flows, *lubrication theory* can be used to derive a model for the flow field based on an ordinary differential equation. In the simplest form (see e.g. Moffatt [23] and Huppert [13]) the approximation assumes that the dependent flow variables change only very slowly in the basic flow direction and that nonlinear effects play no role, so that a separation of variables can be effected, resulting in a parabolic velocity profile in the direction orthogonal to the main flow direction. The width of the stream, however, is an unknown of the problem and is resolved by the boundary conditions applying at the free surface. This type of model is very effective for small Reynolds-number flows, but it models only the gross features one would guess are generated by the underlying geometric features of the flow domain. For example, for the case shown in figure 1, at a Reynolds number of 36.6, the mean-square error between the prediction of the basic lubrication model and the experimental measurements is greater than 48% of the total mean-square deviation of the free surface from its undisturbed depth. At smaller Reynolds numbers the error in the basic lubrication model for flow over the same bumpy bed was reduced, but it nevertheless failed to predict accurately such details as the depth of the free-surface trough between the two bumps at Reynolds numbers above about 5, nor did it predict the undershoot to the right of the two principal wave crests [30].

Including inertial terms in a lubrication model leads to improved predictions. Manton [21] computed the velocity field for a fixed domain whose boundary is slowly varying in the direction of the main flow. In the model proposed for the free-boundary case in [30], the velocity profile may also vary slowly in the direction of the main flow and will not in general remain parabolic. The shape of the profile depends, however, only on a single parameter given by the product of the Reynolds number and the derivative of the fluid thickness at a given point, as found in the fixed-domain calculations by Manton [21]. For the experiment described in the previous section, at Reynolds number 12.2, a lubrication model incorporating such modifications achieved a mean-square accuracy of better than 6%. More importantly, it predicted quite well the depth of the trough between the two wave crests and the undershoot after the two principal waves, as shown in figure 2.

By comparing the lubrication models with laboratory experiments and with finite-element calculations, the improved, nonlinear model has been found to be significantly more accurate, both qualitatively and quantitatively [30], than the standard model which has a fixed velocity profile and does not incorporate inertial effects. However, the range of applicability of the improved model has yet to be studied extensively. For example, it appears not to have a smooth solution for moderate and large Reynolds numbers. It is unknown whether this corresponds in some way to a singularity associated with the physical problem (and presumably the full free-boundary problem) or if it is simply an artifact of the improved lubrication model. Resolution of these questions requires further analyses both of the improved lubrication model and of the full Navier-Stokes free-boundary model.

4. Finite-element models. Finite-element discretizations for free surface flows of the Navier-Stokes equations, are now relatively common. In particular, commercial computer codes are readily available that implement such calculations. In [30], calculations using the FIDAP package [11] are described and the results of those calculations have been compared with our laboratory experiments and with those deriving from the lubrication models. For a range of experiments (of the type described in §2) spanning nearly two orders of magnitude in Reynolds number, viz. for Reynolds numbers lying between 0.37 and 25.5, the mean-square errors ranged from 1.5% to 2.8%. The modest increase in error could be attributed to a variety of effects, including an

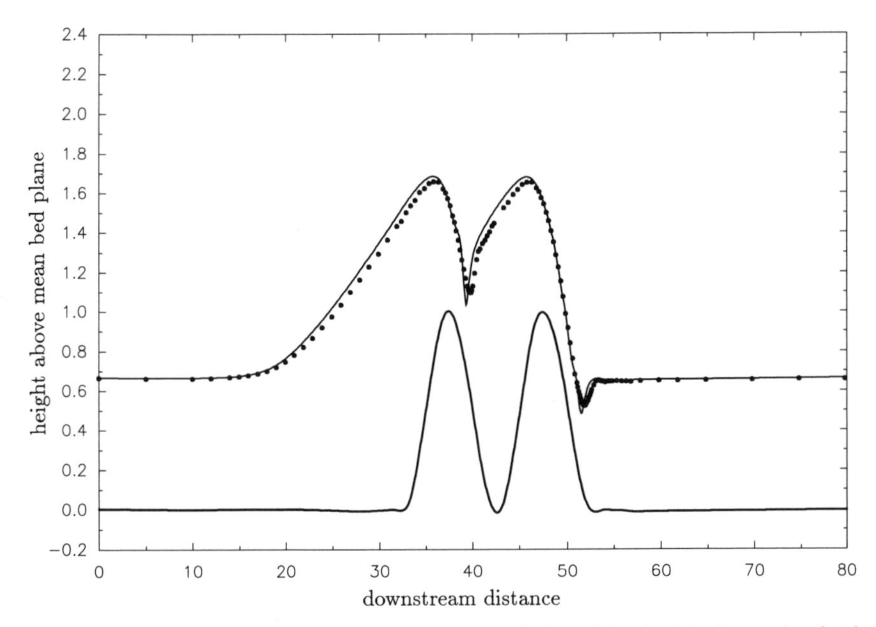

FIG. 2. *Comparison between experimental measurements (indicated by •) of the free-surface height above the plane defined by the uniform part of the channel (the channel bed is denoted by the lower solid curve) and the prediction based on a nonlinear lubrication model (denoted by the upper solid curve). The experimental conditions had Re=12.2, S=3.38 and an upstream depth of 0.664.*

increased three-dimensionality of the flow at the larger Reynolds numbers and a small degradation of data quality as a consequence of an increased complexity (e.g., larger gradients of the free-surface profile) of the flow field. More importantly, all significant features of the flow were well represented by the finite-element calculations, the worst case being shown in figure 3 for the flow at a Reynolds number of 25.5.

The detailed comparison with experiment described in [30], and other experience with finite-element calculations, indicate that such methods can provide extremely accurate and robust models for free-surface flow problems over a wide range of parameters. There are, however, still some shortcomings in existing codes. For example, missing from some commercial codes at the moment is the ability to do bifurcation and sensitivity analysis, as discussed by Scriven at the Workshop. Among the experiments reported in [30], there were two (at Reynolds number 31.4 and 36.6, cf. figure 1) which we were unable to simulate using the finite-element code. We suspect some sort of bifurcation occurs at a Reynolds number somewhere above 25.5, but we have so far been unable to study this feature.

More complicated boundary conditions are also not well treated by some codes. It should be noted that the boundary conditions for the free surface at the upstream and downstream end points of the domain were especially simple for the experimental situation described in §2 and studied in [30]. Away from the bumps, the free surface returns (exponentially fast) to the height it would have on an inclined plane of infinite length having the same angle with respect to the vertical. Thus, the free surface is essentially flat at the points at which it attaches to the inflow and outflow regions of the computational domain. The fluid velocity at the inflow and outflow sections

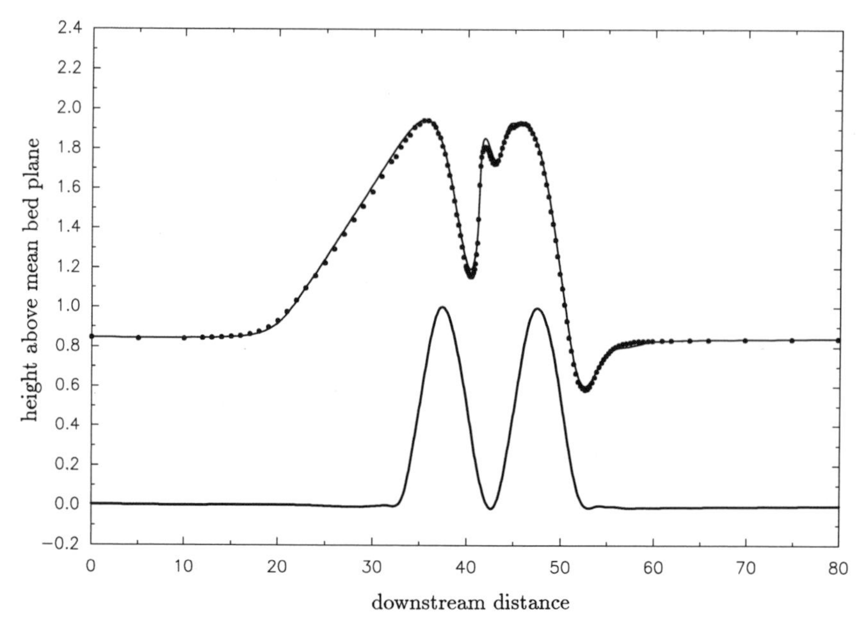

FIG. 3. *Comparison between experimental measurements (indicated by •) of the free-surface height above the plane defined by the uniform part of the channel (the channel bed is denoted by the lower solid curve) and the prediction based on finite element calculations (denoted by the upper solid curve). The experimental conditions had Re=25.5, S=2.08 and an upstream depth of 0.845.*

is a simple parabolic profile in the y-direction and, indeed, this was viewed as a desirable feature in the original design of the experiment. Different kinds of boundary conditions can, however, cause considerable difficulties, but it should be emphasized that some of the problems arise at the mathematical level rather than having to do only with the finite-element discretization. It would seem important, therefore, to develop a better understanding of the role of boundary conditions on the free surface at points of attachment. This will be discussed at greater length in §6, but first we review briefly the existing mathematical theory.

5. Mathematical theory. The mathematical theory for viscous free-surface flows is quite recent. There are different types of problems that have been considered, which fall into either a time-dependent or steady category. The former group is concerned primarily with the evolution of waves on the surface of a fluid of infinite extent or an isolated volume of fluid, for which there is no point of attachment for the free surface. Work in this area can be divided into two groups, depending on whether (see Beale [5] and Allain [2]) or not (see Beale [4], Solonnikov [42] and Sylvester [43]) surface tension is nonzero. (Further references can be found in the papers by Pileckas [27] and Sylvester [43]). It is noteworthy that the inclusion of surface tension in the model leads to stronger regularity properties of the solution, reflecting the physical role of surface tension. More recently, Renardy & Renardy [35] have studied well posedness of various boundary conditions for the time dependent case on finite domains.

In most research on steady-state problems, it is crucial that surface tension be included in the model. The techniques employed all utilize detailed estimates for appropriate fixed-domain problems as developed by Solonnikov & Ščadilov [41] and Nitsche [25]. The free-boundary problem is then approached via an iterative technique based on a contraction mapping principle, which uses in a critical way the property that surface tension is nonzero. In fact, the theories guarantee a solution only for sufficiently small data, and the limit on the magnitude of the data tends to zero with the surface tension parameter. An essential ingredient in the contraction mapping argument is the choice of spaces in which to represent the solution variables.

For fixed-domain problems, there are several choices of function-spaces available for modeling the variables in a viscous-flow problem. Perhaps the simplest are the Hilbert spaces in which the velocity gradient and pressure are square integrable. Certain problems with singularities can not be represented by this class of function space, so more general "Sobolev" spaces, in which the integrability power has been generalized from the integer two to an arbitrary real number, p, satisfying $1 < p < \infty$, have been utilized for existence theories (cf. Serre [39]). These are denoted by W_p^1 for the velocity, with the superscript 1 indicating that derivatives of order 1 (and less) must be p-th power integrable.

In order to demonstrate regularity properties of the solution, function spaces reflecting more localized information are often employed. The most natural spaces to employ would be the spaces, C^k, of k-th power differentiable functions. However, due to mathematical vagaries, boundary-value problems for differential equations are not well posed in such simple spaces. Instead, one must refine the notion of differentiability and consider functions whose derivatives are Hölder continuous, of some order $0 < \alpha < 1$. Such spaces are denoted $C^{k+\alpha}$. It should be noted that the vagaries that lead to this complication are closely related to a prohibition of the use of Sobolev spaces for the natural values of $p = 1$ and $p = \infty$. Other function spaces might be employed as well to demonstrate regularity, such as spaces of analytic functions. Function spaces also often incorporate weights in order to focus on limiting behavior at specified points as employed by e.g. Abergel & Bona [1] and Solonnikov [41].

For free-boundary problems, it has not yet been possible to demonstrate well-posedness in the simplest Hilbert spaces. Jean [15], Pukhnachev [32], Solonnikov [40] and others (cf. the introduction and references cited in [27]) utilized Hölder spaces, with weights to allow for the singularities arising potentially at points of contact between the free surface and solid walls. While this approach has the beneficial effect of establishing both well-posedness and regularity at the same time, it has the disadvantage of requiring more powerful techniques just to get started. The Hilbert-space approach is far simpler to use and additionally allows for a broader range of physical problems to be modeled. Another benefit is that finite-element methods are typically posed in the Hilbert-space setting. Corresponding to the lack of a Hilbert-space theory, there is currently no convergence proof for finite-element models of viscous free-boundary flows.

One reason for casting finite-element methods in a Hilbert-space setting is simplicity. Another is that the displacement functions typically utilized are not sufficiently smooth to be members of the Hölder spaces used in the theoretical developments discussed above. In particular, finite-element spaces of functions that are continuous (C^0) piecewise-polynomials are used routinely to solve flow problems. Requiring spaces of C^1 piecewise-polynomials would be, apparently, unnecessary, and it would complicate computer codes dramatically. On the other hand, the standard finite-element spaces do lie in first-order Sobolev spaces, W_p^1, for any p. Thus it would be natural to attempt to develop a theory for existence, stability and convergence in such spaces following, for example, the methods of Serre [39]. This has been done (see Saavedra & Scott [38]) for a model *scalar* problem, for which an existence theory is first established for the continuous problem. (The ingredients of the complete analysis for the discrete problem will be discussed further in section 9.) However, at the moment there is no analogous theory for the complete free-boundary problem for viscous flows.

6. Boundary conditions. One of the goals of the discussion given in the introductory section was to indicate the crucial importance of the choice of boundary conditions when attempting to model practical flows. We would like now to explore some of these issues in more detail. One particular problem that arises in applications is how boundary conditions at infinity can be transferred to, or simulated by, appropriate conditions on some suitably truncated domain. The die-swell problem described in the introduction provides a very nice example of some of the difficulties that can arise. Let us consider such flows in the specific context of the Stokes equations.

It is usual to suppose that the extrusion process in which the die-swell phenomenon is manifested is of the form indicated in the sketch shown in figure 4a (e.g. see Nickell *et al.* [24], Kruyt *et al.* [19]), in which fluid emanating from the slot of unit thickness forms into a uniform jet of thickness D far away from the exit. Several modeling assumptions have already been made in presuming that the flow is of the form shown in the sketch. One important presumption is that the free surface attaches to the wall at the corners of the slot. There is no physical reason *a priori* why this should obtain in practice, and indeed the experimental evidence is quite to the contrary (see, for example, Jean & Pritchard [16], Tomita & Mochimaru [44] and Pritchard [29]). In one particularly striking example Pritchard [29] gives examples of two steady free-surface flows under exactly the same operating conditions but which are clearly distinguishable from one another by virtue of the fact that they attach to the boundary at different points. In none of the above-cited experiments was the free surface observed to attach to the fixed boundary at a corner and, indeed, there are no experiments known to us in which such a corner attachment has been observed. (We note, in passing, that such observations are not in conflict with the usual assumption for large Reynolds-number flows that flow separation occurs at a sharp corner or a trailing edge. Such an assumption makes good sense for flows at infinite Reynolds numbers, and provides a very good physical model at large Reynolds numbers though, in the latter case there is no theoretical justification to show that the separation takes place precisely at a corner.) Thus, the experimental evidence suggests that, in the absence of any restraining conditions, the point of attachment of the free surface to the boundary can not be specified *a priori* and must therefore be an unknown, to be determined as part of the solution to the flow problem.

Taking the view that the attachment point of the free surface is an unknown of the problem, we must consider the possibility that the attachment could occur at infinity with the fluid filling the entire flow domain available to it. Consider, for the

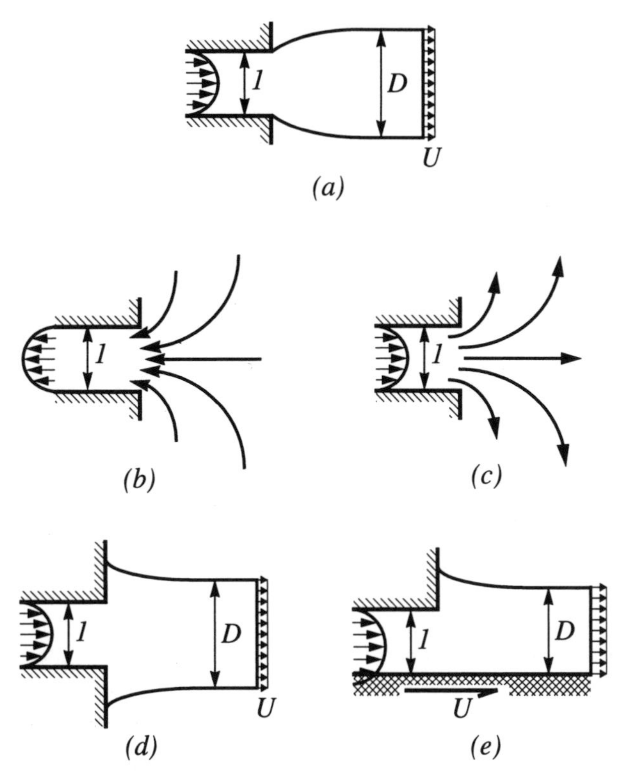

FIG. 4. *Examples of extrusion and coating flows. (a) Extrusion with point of attachment at corner. (b) & (c) Flow out of and into an infinite reservoir (d) Extrusion with an unknown point of attachment. (e) Coating flow with an unknown point of attachment.*

time being, the flow generated by the steady *withdrawal* of fluid through the slot, as indicated in figure 4b, which flow, at large distances from the opening to the slot, closely resembles the classic solution for flow into a sink. Then, because the solution to the Stokes equations is reversed on reversal of the boundary conditions, we see that the kind of flow depicted in figure 4c will provide a solution to the extrusion problem. It is, of course, possible that the latter flow is unstable should it be perturbed, say, by the introduction of a small bubble at the boundary. The situation is apparently even more complicated, in that a one-parameter family of flows seems to be possible. Consider the situations depicted in figures 4d and e, in which the downstream flow has thickness D and a uniform velocity U, where the product UD is constrained to be equal to the volume flow rate per unit width. Then, given any downstream thickness D, there is apparently a solution to the steady flow problem, each of which is presumably distinguished by having its own attachment point on the boundary.

Thus, we see that the actual flow which is realized in an extrusion process can depend sensitively on the the way the final motion is established (i.e. on the resolution of an initial-value problem) or, equivalently on the way the downstream boundary conditions are applied. It would seem that the above considerations are sufficient to account for the quite different experimental realizations of extrusion processes described in the introduction. The above discussion is aimed at illustrating the importance of the choice of boundary conditions in modeling physical problems. However, there are a number of unresolved mathematical issues associated with the choice of boundary conditions for free-surface flows which must be resolved in order to provide a basis for such models.

A mathematical theory for the extrusion problem discussed above does not yet exist in its entirety. Pileckas & Specovius-Neugebauer [26] give an existence theorem for a problem closely related to the flow depicted in figure 4a. The theories of Jean [15] (and see Jean & Pritchard [16]), Pukhnachev [32] and Solonnikov [40] concern a domain of finite extent with the flow required to return to a simple state at the extremities of the flow domain, whereas the flow in the extrusion problem continues unabated indefinitely. The principal modification needed to these theories is their extension to a domain which is unbounded in the flow direction. It may be straightforward to combine the techniques for flows on unbounded domains (see [1] and [26] and references given in [27]) with those developed in [15], [25], [32] and [40] to complete a mathematical theory for the extrusion problem.

An alternative approach to modeling the extrusion problem would be to pose it on a finite domain with appropriate limiting boundary conditions posed not at infinity but at some distant point, e.g. as shown in figures 4a, d & e. A key feature of the industrial free-boundary applications listed in the introduction is the finite extent of the flow domain, leading to a need to formulate appropriate boundary conditions at the end of the flow domain. The finiteness of the flow domain can arise for a number of reasons, including both computational and modeling issues. It may be necessary for computational reasons to truncate the flow domain in order to keep the size of the computational problem manageable. On the other hand, the flow model itself may have a limited domain of applicability, for example, as a result of phase changes in the materials in question, in which case there would be a point at which the free boundary terminates and meets a (real or fictitious) surface.

The theoretical studies of Jean [15], Solonnikov [41] and Pukhnachev [32] have considered flows for which the attachment occurred only at points where the flow velocity is zero. There are several types of boundary conditions that are appropriate for different physical models. We will consider two of them, one of Dirichlet type and the other of Neumann type. Then we will discuss issues relating to regularity properties of the flow field.

6a. *Dirichlet Boundary Conditions for the Free-Surface.* When Dirichlet conditions are posed on the free surface, as might occur, for example, at the endpoints of a computational domain, the slope of the free surface will be unknown there. In order that the flow field be at least continuous as a function of the spatial variables, the flow velocity must be tangential to the free surface at such a point. This poses a potential conflict if Dirichlet data have been given for the velocity without regard for the unknown free-surface slope.

One simple remedy for the above problem might be to relax the Dirichlet condition by adding a component to be determined together with the slope of the free surface. In the (generic) case that the free surface would attach nontangentially to a (possibly fictitious) portion of the domain boundary on which full Dirichlet conditions are posed on the velocity. Moreover, frequently only the component of the velocity normal to this wall is of physical interest. Thus a velocity component tangential to the wall would be able to compensate for any mismatch in the direction of the flow at the point of attachment as depicted in figure 5a. There is no way to specify the *shape* of this velocity component, only its value at the point of attachment; the exact profile presumably could be arbitrary and still yield a well posed problem.

To make this more precise, let a part of the boundary be the segment

$$\Gamma := \{(a,y) \; : \; 0 < y < b\}$$

and suppose that the free surface attaches to Γ at the point $(x,y) = (a,b)$. Let the free boundary be parametrized as the graph of $\gamma(x)$. We are therefore supposing that the boundary condition on γ at the point of attachment is of Dirichlet type, that is, $\gamma(a) = b$. Let $\mathbf{u} = (u,v)$ denote the velocity. Then an appropriate Dirichlet-type boundary condition on the velocity might be of the form (see figure 5a)

$$(1) \quad u(a,y) = (\mathbf{u}\cdot\mathbf{n})\,(a,y) = g(y), \quad v(a,y) = (\mathbf{u}\cdot\mathbf{t})\,(a,y) = \alpha f(y), \quad 0 < y < b,$$

where α will be determined as part of the unknowns in such a way that the flow is tangential to the free surface, that is,

$$(2) \qquad \alpha = \frac{g(b)\gamma'(b)}{f(b)} \,.$$

Note that $g(b) \neq 0$ requires that $f(b) \neq 0$ since we cannot be assured that we would always have $\gamma'(b) = 0$. Conditions (1), with α as in (2), lead to consistent boundary values and presumably a continuous velocity field. Although no theory exists for such boundary-value problems, computational evidence, based on a code we have developed [9], suggests that (1)-(2) provide a valid set of boundary conditions.

On the other hand, one could instead pose only partial Dirichlet conditions for the velocity field on such portions of the boundary. For example, the normal component might be specified, with the tangential component left free. In a variational formulation, this would amount to specifying a natural boundary condition on that portion of the boundary, e.g. that the shear stress vanish. One question is whether such an approach would actually lead to a smooth solution. For example, would the resulting velocity field be tangential to the free surface at the point of attachment?

6b. *Neumann Boundary Conditions for the Free-Surface.* If Neumann conditions are posed on the free surface, then the flow direction is known at the point of attachment. However, the point of attachment itself, and therefore the extent of the domain, is then unknown. Thus, standard Dirichlet conditions on the flow velocity do not make sense. Some sort of stretching must be included in the formulation of the boundary-value problem in addition to, say, a prescribed velocity profile. Again, no theory exists regarding the well posedness of such an approach.

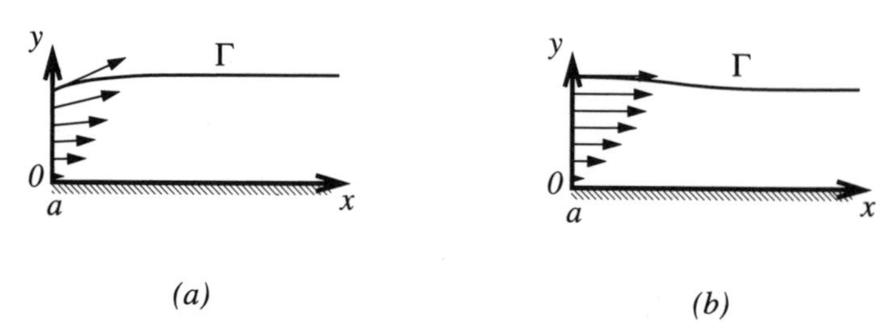

FIG. 5. *Examples of Dirichlet (a) and Neumann (b) data for the free-surface and the corresponding prescription of the velocity field. In (b) the prescribed slope θ of the free surface is zero.*

As an example, suppose that the boundary condition on γ is $\gamma'(a) = \theta$. Then an appropriate Dirichlet-type boundary condition on Γ for the velocity might be of the form

$$\mathbf{u} = \mathbf{f}(y/\gamma(a))/\gamma(a), \quad 0 < y < \gamma(a), \tag{3}$$

where $\mathbf{f}$ is an arbitrary function defined on $[0,1]$ (see figure 5b, in which a flow is depicted having $\theta = 0$). For example, in the problem of flow down a flat plate as discussed in §2, $\mathbf{f}(y) := \left(\frac{3}{2}y(2-y), 0\right)$. Here we have picked a particularly simple scaling for the data. Note that the flux

$$\int_\Gamma \mathbf{u}\cdot\mathbf{n}\,ds = \int_0^{\gamma(a)} \mathbf{f}(y/\gamma(a))\cdot\mathbf{n}/\gamma(a)\,dy = \int_0^1 \mathbf{f}(\eta)\cdot\mathbf{n}\,d\eta$$

is independent of $\gamma(a)$.

One might also be interested in more general scalings of the form

$$\mathbf{u} = \mathbf{f}(\beta(y,\gamma(a)))\beta_y(y,\gamma(a)), \quad 0 < y < \gamma(a), \tag{4}$$

where $y \mapsto \beta(y,t)$ is an invertible mapping of the interval $0 < y < t$ to the interval $[0,1]$ for all $t > 0$ and β_y denotes the partial derivative of β with respect to its first argument. With such a scaling, we still have

$$\int_\Gamma \mathbf{u}\cdot\mathbf{n}\,ds = \int_0^{\gamma(a)} \mathbf{f}(\beta(y,\gamma(a)))\cdot\mathbf{n}\ \beta_y(y,\gamma(a))\,dy = \int_0^1 \mathbf{f}(\eta)\cdot\mathbf{n}\,d\eta,$$

so that flux equality is maintained independently of the unknown $\gamma(a)$. It is not known if this type of Dirichlet problem leads to a well posed problem under appropriate restrictions on $\mathbf{f}$ and β, but numerical experiments using our own code [9] suggest that it does work well.

6c. *Regularity of the Flow Field.* The above discussion has focused only on the simplest compatibility conditions that would lead to continuity of the velocity at points of attachment of the free boundary. It is unclear whether higher-order compatibility conditions are required to guarantee that solutions be smooth unless, by chance, fortuitous data were posed for the problem. For example, if one were able to guess correctly the *exact* form of the asymptotic solution of the free-boundary problem upstream and downstream of the computational region, then smoothness could be guaranteed. The resulting domains, having a boundary given in part by the free-surface, will typically have angular points at which we can expect mild singularities in the velocity (cf. Blum & Rannacher [8]) unless the angle of attachment of the free surface is quite special. This observation also holds for the boundary-value problem in which the normal component of the velocity is specified together with a vanishing shear stress (as a natural boundary condition). The asymptotic solution cannot be expected to have the shear stress exactly zero at some point not at infinity.

On the other hand, it is conceivable that the free surface would adjust itself appropriately in all cases to provide a smooth solution for the given data once the first compatibility condition is satisfied. Although the Dirichlet data might not match exactly the asymptotic form of the smooth solution, the slope of the free surface might also differ slightly in such a way that a nearby smooth solution is obtained. Investigation of the smoothness of such finite-domain boundary-value problems for free surfaces would be of significant interest.

7. Moving contact lines. A much more difficult boundary-value problem occurs with regard to moving contact lines, as discussed by Dussan, Hassager, Scriven and others at the Workshop. In this case, there is an imposed discontinuity in the velocity field at the point where the free surface contacts a solid boundary. For example, suppose that the fluid contacts the solid wall on the infinite segment $\Gamma := \{(a,y) : y < b\}$ and suppose that the free surface (parametrized as the graph of $\gamma(x)$) attaches at the point (a,b), so that the boundary condition on γ at the point of attachment is of Dirichlet type, that is, $\gamma(a) = b$. Let $u = (u,v)$ denote the velocity. Then an appropriate Dirichlet-type boundary condition on the velocity might be of the following form:

$$u(a,y) = (\mathbf{u}\cdot\mathbf{n})(a,y) = 0, \quad v(a,y) = (\mathbf{u}\cdot\mathbf{t})(a,y) = -1, \tag{5}$$

for some interval of y including b. If the angle of attachment is nontangential to Γ, then the fact that the free surface must be a streamline forces $\mathbf{u}$ to be discontinuous at (a,b).

It is possible to have an existence theory for the Navier-Stokes equations in spaces that allow such discontinuities. The Sobolev spaces, W_p^1, consisting of functions whose gradients are p-th power integrable, include discontinuous functions when $p < n$ (where n is the dimension of the spatial domain, $n = 2$ in the examples here). It has been shown by Serre [39] that the Navier-Stokes equations are well posed in such spaces for sufficiently smooth domains. However, for the moving-contact-line problem, it is not clear whether the free boundary would possess the appropriate amount of smoothness required for the theory of [39]. There may well be a subtle interplay between the regularity of the free boundary and the resulting flow field.

8. Surface tension effects. As observed in the introduction regarding the die-swell problem, surface tension can play a significant role from a physical point of view. Its role in the existing mathematical theories was discussed in section 5, as well. Here we focus on the importance of surface tension in the well posedness of numerical algorithms. Not surprisingly, it has been of considerable importance, possibly reflecting its physical role and the apparent lack of regularity associated with the mathematical models in its absence.

Many computer codes make the assumption that surface tension is nonzero in the underlying algorithms. Moreover, computational experience indicates that the difficulty of isolating solutions increases dramatically as surface tension is decreased. This is a parameter that can be varied (in a nonphysical way) in order to continue a more easily obtained solution (for large surface tension) to one having the desired value of surface tension. Thus, it becomes a question of significant interest whether a well posed model exists for the viscous free-boundary problem in the absence of surface tension. As far as we know, it remains an open problem to demonstrate the existence of solutions in this case, although a closely related problem has been solved by Bemelmans [6] in which the model is augmented by the addition of a force of self-attraction for the fluid. We also note that the results of Solonnikov [42] for the time-dependent evolution of an isolated volume of fluid yield steady solutions in the limit as time approaches infinity.

Regarding numerical algorithms, the absence of surface tension also causes difficulties. When surface tension is nonzero, there is a jump in the normal stress across the free surface which is equal to a constant (the surface tension) times the curvature of the free surface. The curvature operator is elliptic (albeit nonlinear), and this property allows one to formulate approaches to determine the free-surface location in a stable way. Without surface tension, such a relationship is lost, and one must find some other way of determining the free-surface location.

One approach that has been described [24] makes use of the fact that the free surface (in two dimensions) must be a streamline. Thus one can parameterize the free surface as $\{\mathbf{x}(s)\}$ such that

$$\frac{d\mathbf{x}}{ds} = \mathbf{u}(\mathbf{x}(s)) \ . \tag{6}$$

Typically, this is augmented by a known point of attachment: $\mathbf{x}(a) = \mathbf{x}^0$. While this approach seems appealing, it is plagued by the fact that, at a point of attachment, $\mathbf{u}(\mathbf{x}^0) = \mathbf{0}$, hence there is no well defined way to leave the point $\mathbf{x}^0$. That is, $\mathbf{x}(s) = \mathbf{x}^0$ is a solution for all s, and any other solution emanating from that point would be problematical. The same (constant) solution is generated by any explicit difference scheme used to solve (6).

Undaunted by this obstacle, practitioners have leapt over the gap in information by using an implicit time-stepping scheme, and one then apparently obtains a solution different from $\mathbf{x}^0$ for the first time step and subsequent time steps. How this comes about is not entirely straightforward. If one considers, say, the backward Euler scheme

$$\frac{\mathbf{x}^1 - \mathbf{x}^0}{\Delta s} = \mathbf{u}(\mathbf{x}^1),$$

then Taylor's theorem yields

$$\left|\frac{\mathbf{x}^1 - \mathbf{x}^0}{\Delta s}\right| = \left|\mathbf{u}(\mathbf{x}^1)\right| = \left|\mathbf{u}(\mathbf{x}^1) - \mathbf{u}(\mathbf{x}^0)\right| \leq C\left|\mathbf{x}^1 - \mathbf{x}^0\right|.$$

Assuming $\mathbf{x}^1 \neq \mathbf{x}^0$ (as we would want) then we get the disconcerting conclusion that $\Delta s \geq 1/C$. Thus one cannot obtain a converged solution by taking smaller and smaller Δs. The meaning of solutions obtained for large Δs is questionable. Other implicit time-stepping schemes would of course suffer the same dilemma.

We note that the recent approach in [10] is different and may avoid these difficulties.

9. Convergence estimates. Even for free-boundary problems having a well established mathematical theory and demonstrated success in finite-element calculations, there is little in the way of convergence analysis for the finite-element discretization. No simple variational formulation in conventional Hilbert spaces has yet appeared (see the discussions in §5 and [30] for details) in which finite-element convergence can be established by traditional means. Research regarding convergence has been initiated by Saavedra & Scott [38] who studied a related scalar problem in a nonstandard framework related to that used by Serre [39]. In particular, existence for the (scalar) field variable is established in W_p^1 for appropriate $p > 2$ and for the free-boundary in the class of Lipschitz functions (which is the same as the Sobolev space W_∞^1). As observed in §5, no such theory exists for the full flow equations with free-boundary.

Following closely the existence theory for the continuous problem, the existence and stability of the corresponding discrete solution is easily established, as described in [38]. The key ingredient in the proof is a new stability bound for the finite element discretization in W_p^1 for (fixed) Lipschitz domains. (By a simple change of coordinates, such problems are equivalent to ones in more regular domains but having only bounded, measurable coefficients.) Such estimates are proved using the stability properties of the finite element method in the W_∞^1 norm established by Rannacher & Scott [34] for more regular domains, together with a perturbation argument. Corresponding stability estimates (e.g. for the velocity approximation in W_∞^1) for the finite element approximation of the Navier-Stokes equations (even in fixed domains) do not appear to be known at the moment.

Once stability of the discrete approximation has been established, convergence estimates with respect to mesh refinement are established in [38] using a standard consistency analysis (in a finite-element setting). Optimal-order convergence results are proved in [38] for problems having Dirichlet boundary conditions for the free surface (cf. §6a). However, the same consistency analysis applied to a corresponding problem with Neumann boundary conditions for the free surface (cf. §6b) seems to imply that the discrete approximation is not fully consistent. As a result it is not at all clear that the convergence rate would be of optimal order in the case of Neumann boundary conditions for the free surface, and no convergence estimates were proved in [38] for that case. In some recent numerical studies by Juarez and Saavedra [18], it has been found that optimal-order convergence also apparently holds in the case of Neumann conditions for the free surface. However, a rigorous convergence analysis in this case is lacking.

10. Conclusions. Free-boundary problems offer a rich set of interesting theoretical problems of significant practical, industrial interest. These span a broad range of mathematical areas, including bifurcation theory, existence theory for partial differential equations, and numerical analysis. Advances in these areas should lead to improvements in technology for the design and control of important industrial processes.

11. Acknowledgements. The work of Scott was supported in part by the National Science Foundation through award number DMS-8903548. The work of Pritchard and Tavener was supported in part by the National Science Foundation through award number DMS-8805311-04.

REFERENCES

[1] F. Abergel and J. L. Bona, A mathematical theory for viscous, free-surface flows over a perturbed plane, Report no. AM81 (1991), Dept. Math., Penn State Univ. To appear in *Arch. Rat. Mech. & Anal.*

[2] G. Allain, Small-time existence for the Navier-Stokes equations with a free surface, *Applied Math. and Optimization* **16** (1987), 37-50.

[3] J. Batchelor, J. P. Berry and F. Horsfall, Die swell in elastic and viscous fluids, *Polymer* **14** (1973), 297-299.

[4] J. T. Beale, The initial value problem for the Navier-Stokes equations with a free surface, *Comm. Pure & Appl. Math.* **XXXIV** (1981), 359-392.

[5] J. T. Beale, Large-time regularity of viscous surface waves, *Arch. Rat. Mech. & Anal.* **84** (1984), 307-352.

[6] J. Bemelmans, On a free boundary for problem the Navier-Stokes equations, *Ann. Inst. H. Poincaré Anal. Nonlin.* **4** (1987), 517-547.

[7] T. B. Benjamin, W. G. Pritchard and S. J. Tavener, Steady and unsteady flows of a highly viscous liquid inside a rotating horizontal cylinder, in preparation.

[8] H. Blum and R. Rannacher, On the boundary value problem of the biharmonic operator on domains with angular corners, *Math. Meth. Appl. Sci.* **2** (1980) 556-581.

[9] S. Brenner, W. G. Pritchard, L. R. Scott and S. J. Tavener, The nonconforming Crouzeix-Raviart element for computation of free-surface flows, in preparation.

[10] J. Descloux, R. Frosio and M. Flück, A two fluids stationary free boundary problem, *Comp. Meth. Appl. Mech. & Eng.* **77** (1989), 215-226.

[11] M. S. Engelman, FIDAP — A fluid dynamics analysis package, *Adv. Eng. Software* **4** (1982), 163-.

[12] S. L. Goren and S. Wronski, The shape of low speed capillary jets of Newtonian liquids, *J. Fluid Mech.* **25** (1966), 185-198.

[13] H. E. HUPPERT, The propagation of two–dimensional and axisymmetric viscous gravity currents over a rigid horizontal surface, *J. Fluid Mech.* **121** (1982), 43–58.

[14] H. E. HUPPERT, The intrusion of fluid mechanics into geology, *J. Fluid Mech.* **173** (1982), 557–594.

[15] M. JEAN, Free surface of the steady flow of a Newtonian fluid in a finite channel, *Arch. Rat. Mech. & Anal.* **74** (1980), 197–217.

[16] M. JEAN AND W. G. PRITCHARD, The flow of fluids from nozzles at small Reynolds number, *Proc. Roy. Soc. Lond. Ser. A* **370** (1980), 61–72.

[17] D. D. JOSEPH, J. NELSON, M. RENARDY AND Y. RENARDY, Two–dimensional cusped interfaces, *J. Fluid Mech.* **223** (1991), 383–409.

[18] L. H. JUÁREZ AND P. SAAVEDRA, Numerical solution of a model free–boundary problem, preprint.

[19] N. P. KRUYT, C. CUVELIER, A. SEGAL AND J. VAN DER ZANDEN, A total linearization method for solving viscous free boundary flow problems by the finite element method. *Int. J. Numer. Meth. Fluids* **8** (1988), 351–363.

[20] A. S. LODGE, *Elastic liquids, and introductory vector treatment of finite–strain rheology*, London; New York: Academic Press, 1964.

[21] M. J. MANTON, Low Reynolds number flow in slowly varying axisymmetric tubes, *J. Fluid Mech.* **49** (1971), 451–459.

[22] S. MIDDELMAN AND J. GAVIS, Expansion and contraction of capillary jets of Newtonian liquids, *Physics of Fluids* **4** (1961), 355–359.

[23] H. K. MOFFATT, Behaviour of a viscous film on the outer surface of a rotating cylinder, *J. de Mécanique* **16** (1977), 651–673.

[24] R. E. NICKELL, R. I. TANNER AND B. CASWELL, The solution of viscous incompressible jet and free–surface flows using finite–element methods, *J. Fluid Mech.* **65** (1974), 189–206.

[25] J. A. NITSCHE, Free boundary problems for Stokes' flows and finite element methods, *Equadiff* **6**, *Lecture Notes in Math.* **1192**, Berlin: Springer–Verlag, 1986, 327–332.

[26] K. PILECKAS AND M. SPECOVIUS-NEUGEBAUER, Solvability of a problem with free noncompact boundary for a stationary Navier-Stokes system. I, *Lithuanian Math. J.* **29** (1990), 281–292.

[27] K. PILECKAS AND W. M. ZAJACZKOWSKI, On the free boundary problem for stationary compressible Navier-Stokes equations, *Commun. Math. Phys.* **129** (1990), 169–204.

[28] W. G. PRITCHARD, Some viscous–dominated flows, *Trends and Applications of Pure Mathematics to Mechanics*, P. G. Ciarlet and M. Roseau, eds., *Lecture Notes in Physics* **195**, Berlin: Springer–Verlag, 1984, 305–332.

[29] W. G. PRITCHARD, Instability and chaotic behaviour in a free–surface flow, *J. Fluid Mech.* **165** (1986), 1–60.

[30] W. G. PRITCHARD, L. R. SCOTT AND S. J. TAVENER, Viscous free–surface flow over a perturbed inclined plane, Report no. AM83 (1991), Dept. Math., Penn State Univ., *Philos. Trans. Roy. Soc. London*, submitted.

[31] V. V. PUKHNACHEV, Invariant solutions of the Navier–Stokes equations describing the motion of a free boundary, *Doklady Nauk Akademii SSSR* **202** (1972), 302–305 (*in Russian*).

[32] V. V. PUKHNACHEV, Hydrodynamic free boundary problems, *Nonlinear Partial Differential Equations and their Applications, Collège de France Seminar Volume III*, Boston: Pitman, 1982, 301–308.

[33] M. Rabaud, S. Michelland and Y. Couder, Dynamical regimes of directional viscous fingering: spatiotemporal chaos and wave propagation, *Phys. Rev. Lett.* **64** (1990), 184–187.

[34] R. Rannacher and L. R. Scott, Some optimal error estimate for piecewise linear finite element approximations, *Math. Comp.* **38** (1982), 437–445.

[35] M. Renardy and Y. Renardy, On the nature of boundary conditions for moving free surfaces, *J. Comp. Phys.* **93** (1991), 325–355.

[36] S. Richardson, Two–dimensional bubbles in slow viscous flows, *J. Fluid Mech.* **33** (1968), 475–493.

[37] S. Richardson, Two–dimensional bubbles in slow viscous flows. Part 2., *J. Fluid Mech.* **58** (1973), 115–128.

[38] P. Saavedra and L. R. Scott, Variational formulation of a model free–boundary problem, *Math. Comp.* (1991), to appear.

[39] D. Sèrre, Equations de Navier–Stokes stationnaires avec données peu régu-lières, *Annali della Scuola Normale Superiore di Pisa, serie IV,* **X** (1983), 543–559.

[40] V. A. Solonnikov and V. E. Ščadilov, On a boundary value problem for a stationary system of Navier–Stokes equations, *Proc. Steklov Inst. Math.* **125** (1973), 186–199.

[41] V. A. Solonnikov, Solvability of a problem in the plane motion of a heavy viscous incompressible capillary liquid partially filling a container, *Math. USSR Izvestia* **14** (1980), 193–221.

[42] V. A. Solonnikov, On the transient motion of an isolated volume of a viscous incompressible fluid, *Math. USSR Izvestia* **31** (1988), 381–405.

[43] D. L. G. Sylvester, Large time existence of small viscous surface waves without surface tension, *Commun. P. D. E.* **15** (1990), 823–903.

[44] Y. Tomita and Y. Mochimaru, Normal stress measurements of dilute polymer solutions, *J. Non–Newtonian Fluid Mech.* **7** (1980), 237–255.

HIGH ORDER BOUNDARY INTEGRAL METHODS FOR VISCOUS FREE SURFACE FLOWS

J.J.L. HIGDON* AND C.A. SCHNEPPER*

Abstract. This paper describes the use of high order boundary integral techniques for viscous free surface flows. Two approaches are discussed. The first is a coordinate based algorithm using spherical coordinates with up to 4th order representation. The second approach is based on a spectral element formulation allowing arbitrary order approximation with orthogonal polynomials in mapped coordinates.

1. Introduction. Free surface flows are among the most common flows encountered in nature and in engineering processes. These flows span a wide range of Reynolds numbers from the Stokes flow of microscopic droplets to the nearly inviscid motion of high Reynolds number water waves. The most unique feature of free surface flows is the fact that the boundaries of the fluid are not determined a priori, but rather are determined by the dynamics of the system. This feature is a major source of difficulty for those who would attempt to model such flows theoretically. In numerical computations, one cannot rely on a simple grid or element structure, but must continually adapt these representations to the shape of the fluid domain. A further complication arising in free surface flows is the specification of boundary conditions at the surface. Instead of simple no-slip or specified velocity boundary conditions, we require conditions on the continuity of stress and velocity across an interface. With the addition of surface tension, a jump in normal and/or shear stress arises. These boundary conditions place restrictions on the type of algorithm which may be used to solve such boundary value problems. Finally, and most importantly, we note that the specification of the boundary conditions on a surface of unknown position creates a non-linear problem for all free surface flows, even for those where the governing equations are themselves linear.

In flow at low Reynolds number, which is the subject of the present effort, we are concerned with the dynamics of microscopic droplets and bubbles, as well as the stability and flow of liquid films. For researchers working on this regime, the boundary integral method has proved to be the overwhelming favorite for numerical computations. The first application of this technique to viscous free surface flows was by Youngren and Acrivos [26]. These authors studied the deformation of inviscid bubbles in an axisymmetric viscous straining field for arbitrary capillary numbers. Rallison and Acrivos [22] performed similar calculations for viscous drops in axisymmetric flows, while Rallison [20] examined the deformation of viscous drops in plane shear flow. This last work is noteworthy for solving the full three dimensional flow problem, albeit with a coarse discretization and limited parameter range. A further discussion of research on the deformation and break-up of viscous droplets is given by Rallison [21]. Leal and coworkers [4,11,24] have studied the motion of particles and droplets relative to deformable interfaces and have

*Department of Chemical Engineering, University of Illinois, Urbana, Illinois 61801. This work was supported by the National Science Foundation.

examined transient effects on the break-up of elongated droplets in straining fields. Pozrikidis [13,16-18] has used boundary integral computations for modelling the large scale deformation of droplets rising through a quiescent fluid and for liquid layers adjacent to rigid walls.

In our research program, we have employed the boundary integral technique in a variety of applications. Higdon [5,6] studied Stokes flow over two dimensional cavities and obstacles. Larson and Higdon [9,10] studied the flow near the boundary of a porous medium using the boundary integral technique both for the Stokes equations and for Laplace's equation. Schnepper [23] developed a high order algorithm for modelling the deformation of drops and bubbles in viscous flows. Muldowney [12] developed a spectral form of the boundary integral method as part of a more general spectral element algorithm. Cohn [3] and Occhialini [14] have used similar spectral implementations for Laplacian, Helmholtz and Stokes equations.

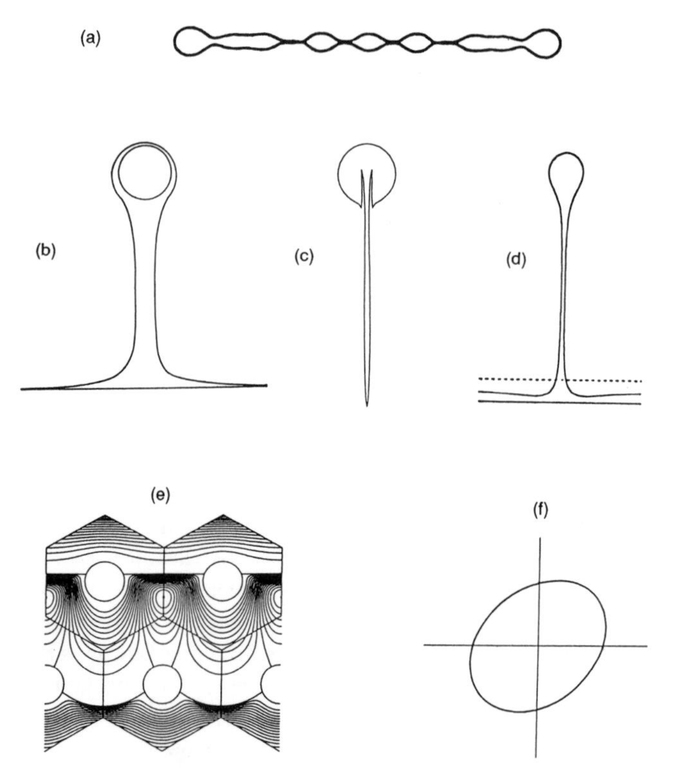

Figure 1 Applications of the boundary integral method in Stokes flow. (a) relaxation instability of an extended droplet [24], (b) a rigid particle moving through a deformable interface [4], (c) the instability of a buoyant droplet [17], (d) a buoyant liquid layer detaching from a rigid wall, (e) the complex streamlines near the boundary of a porous medium [10], (f) the equilibrium state of a viscous droplet in a simple shear flow – a three dimensional flow field.

A selection of results from the references above is shown in Figure 1. From the breadth and complexity of these flows, one may conclude the two dimensional and axisymmetric Stokes flows may readily be calculated with existing boundary integral techniques. By contrast, for three dimensional flows, studies have been limited to spherical or nearly spherical bodies with a large computational effort required even for these simple geometries. In the present paper, we address the computational challenge of calculating three dimensional viscous free surface flows.

2. Boundary integral formulation. The flow of an incompressible viscous fluid at low Reynolds number is governed by the Stokes equations and the continuity equation

$$-\nabla p + \mu \nabla^2 \boldsymbol{u} = 0 \tag{1}$$

$$\nabla \cdot \boldsymbol{u} = 0 \tag{2}$$

In the Stokes equations, p refers to the dynamic pressure, hence the total pressure is $P = p + \rho g z$. For multiphase flows, the governing equations (1,2) apply in each phase with respective density ρ_i and viscosity μ_i.

Let S_B be an interface between two fluids and define the surface stress $\boldsymbol{f} = \boldsymbol{\sigma} \cdot \boldsymbol{n}$. The interfacial boundary conditions for velocity and surface stress are then

$$\boldsymbol{u}_1 = \boldsymbol{u}_2 \tag{3}$$

$$\boldsymbol{f}_1 - \boldsymbol{f}_2 = \gamma(\mathrm{div}\,\boldsymbol{n})\boldsymbol{n} \tag{4}$$

In the stress boundary condition (4), the term on the right is the jump in normal stress owing to interfacial tension γ. In these terms the stress $\boldsymbol{f}$ is based on the dynamic pressure p. For two fluids of different density, an additional term $(\rho_2 - \rho_1)gz$ appears corresponding to the jump in hydrostatic pressure.

With the introduction of the fundamental solution of the Stokes equations, the partial differential equations and boundary conditions may be converted into a system of integral equation over the boundaries and interfaces of the fluid domains. The numerical solution of these integral equations is referred to as the boundary integral method (also called the boundary element method). A brief derivation of these integral equations is given by Higdon, while more extensive expositions may be found in the recent books by Kim and Karrila [8] and by Pozrikidis [19]. In the following, we give only the essential results.

The fundamental solution of the Stokes equations S_{ij} is called the Stokeslet or the Oseen tensor. Physically $S_{ij}(\boldsymbol{x} - \boldsymbol{x}_0) f_j / 8\pi\mu$ represents the velocity u_i at a point $\boldsymbol{x}_0$ induced by a point force of strength f_j located at the point $\boldsymbol{x}$. The stress σ_{ik} associated with the point force is expressed $T_{ijk} f_j / 8\pi\mu$. For the three dimensional Stokes equations, S_{ij} and T_{ijk} are given by

$$S_{ij} = \frac{\delta_{ij}}{r} + \frac{\widehat{x}_i \widehat{x}_j}{r^3} \qquad T_{ijk} = -6\mu \frac{\widehat{x}_i \widehat{x}_j \widehat{x}_k}{r^5} \tag{5}$$

where $\widehat{\boldsymbol{x}} = \boldsymbol{x} - \boldsymbol{x}_0$ and $r = |\widehat{\boldsymbol{x}}|$.

With these definitions, the fundamental integral formula for the Stokes equations may be written

$$u_i(\boldsymbol{x_0}) = \frac{1}{8\pi\mu} \int\limits_{S_B} [S_{ij}(\widehat{\boldsymbol{x}})f_j - T_{ijk}(\widehat{\boldsymbol{x}})u_j n_k] dS \tag{6}$$

This integral formula states that the velocity at any point in the *interior* of the fluid domain may be expressed as an integral of the fluid velocity and the surface stress over the boundary of the domain. For a point on the *boundary* of the domain, the same equation applies with a factor of $1/4\pi\mu$ in front of the integral. The terms associated with S_{ij} and T_{ijk} are referred to as *single layer* and *double layer* distributions respectively. It should be noted that some authors employ alternative definitions of S_{ij} and T_{ijk} absorbing the factor $(1/8\pi\mu)$ into each term.

While the integral formula (6) is the fundamental result on which all boundary integral methods are based, there are several important variations of this result which are useful in different contexts. For *exterior* flows about *rigid bodies*, integral formulas based on (6) may be written for the actual flow and for the undisturbed flow field $\boldsymbol{u}^\infty$ in the absence of the particle. Subtracting the two equations, and applying the no-slip boundary condition yields for a point $\boldsymbol{x_0}$ on the *surface* of the particle

$$-u_i(\boldsymbol{x_0}) = \frac{1}{8\pi\mu} \int\limits_{S_B} [S_{ij}(\widehat{\boldsymbol{x}})f_j] dS \tag{7}$$

This formulation has the advantage that it requires the evaluation of only one kernel, and it was the method of choice for much of the early work using the boundary integral method. Unfortunately, this integral equation of the *first kind* may lead to an ill-conditioned system of equations. In an alternative approach, Power and Miranda [15] show that the single layer distribution may be eliminated by adding additional singularities in the interior of the body. This leads to an equation of the *second kind*. With further modifications, Kim and Karrila show that this *completed double layer* formulation may be used as the basis for an efficient iterative solution of a well conditioned set of integral equations.

For viscous free surface flow, the variable of interest is the fluid velocity evaluated at the free surface. For clarity, consider the case of a droplet of viscosity $\mu_1 = \lambda\mu$ immersed in a fluid of viscosity $\mu_2 = \mu$ with undisturbed velocity $\boldsymbol{u}^\infty$. An integral formula (6) is written for each phase with the point $\boldsymbol{x_o}$ on the interface. Subtracting the two expressions and employing the boundary conditions (3,4) leads to

$$\begin{aligned} u_i(\boldsymbol{x_0}) = {} & \frac{2}{1+\lambda}\, \boldsymbol{u}_i^\infty(\boldsymbol{x_0}) - \left(\frac{\gamma}{1+\lambda}\right) \frac{1}{4\pi\mu} \int\limits_{S_B} [S_{ij}(\widehat{\boldsymbol{x}})(\operatorname{div}\boldsymbol{n})n_j] dS \\ & - \left(\frac{\lambda-1}{\lambda+1}\right) \frac{1}{4\pi\mu} \int\limits_{S_B} T_{ijk}(\widehat{\boldsymbol{x}})u_j n_k] dS \end{aligned} \tag{8}$$

This is an integral equation of the second kind for the velocity $\boldsymbol{u}$. The integral of S_{ij} involves only the known functions $\gamma(\text{div}\,\boldsymbol{n})n_j$ representing the jump in normal stress. If the fluids had different densities, this term would be augmented by a jump in a hydrostatic pressure as discussed previously. The integral equation (8) may be solved via a convergent Picard iteration for all values of λ in the range $0 < \lambda < \infty$. This iterative solution technique has been utilized by Pozrikidis [18]. In the limiting cases, $\lambda = 0,\ \infty$; the integral operator has an eigenvalue equal to 1, and the iteration fails to converge.

For unsteady flows, the surface velocity (8) combined with the kinematic boundary condition gives the condition for updating the position of the free surface. For steady flows, the linear integral equation (8) combined with the kinematic boundary condition yields a non-linear equation for the position of the free surface. If the position of the free surface is described by some function a or by a vector of parameters $\boldsymbol{a}$, we seek the solution $\boldsymbol{a}$ for which $u_n \equiv \boldsymbol{u} \bullet \boldsymbol{n} = 0$ at the interface.

The standard algorithm for solving non-linear systems is Newton's method which requires the calculation of the Jacobian matrix $\partial u_n(\boldsymbol{x_0})/\partial \boldsymbol{a}$. Let us suppose that the discretization of the problem leads to N surface nodes $\boldsymbol{x_0}$ and N degrees of freedom $\boldsymbol{a}$. The question arises as to how one calculates the Jacobian matrix. One approach is to differentiate numerically by perturbing each of the parameters $\boldsymbol{a}$ in turn. This requires the solution of N boundary integral equations, each with a *different geometry*. Thus all integrals must be reevaluated and a different matrix must be inverted each time. A more reasonable alternative is to calculate the Jacobian by direct differentiation of the integral equation. For simplicity, let us rewrite (8) in short form with all the constants indicated by *

$$u_i(\boldsymbol{x_0}) = *u_i^\infty + * \int_{S_B} [S_{ij}(\text{div}\,\boldsymbol{n})n_j]dS + * \int_{S_B} [T_{ijk}u_j n_k]dS \tag{9}$$

Differentiate with respect to the parameters $\boldsymbol{a}$

$$\frac{\partial u_i}{\partial \boldsymbol{a}} = *\frac{\partial u_i^\infty}{\partial \boldsymbol{a}} + *\frac{\partial}{\partial \boldsymbol{a}} \int_{S_B} [S_{ij}(\text{div}\,\boldsymbol{n})n_j]dS + *\frac{\partial}{\partial \boldsymbol{a}} \int_{S_B} [T_{ijk}u_j n_k]dS \tag{10}$$

Now carry through the differentiation of each integral in turn

$$\frac{\partial}{\partial \boldsymbol{a}} \int_{S_B} S_{ij}(\text{div}\,\boldsymbol{n})n_j dS = \int_{S_B} \frac{\partial S_{ij}}{\partial \widehat{x}_l}\frac{\partial \widehat{x}_l}{\partial \boldsymbol{a}}(\text{div}\,\boldsymbol{n})n_j dS + \int_{S_B} S_{ij}\frac{\partial}{\partial \boldsymbol{a}}[(\text{div}\,\boldsymbol{n})n_j dS] \tag{11}$$

$$\begin{aligned} \frac{\partial}{\partial \boldsymbol{a}} \int_{S_B} T_{ijk}u_j n_k dS = &\int_{S_B} \frac{\partial T_{ijk}}{\partial \widehat{x}_l}\frac{\partial \widehat{x}_l}{\partial \boldsymbol{a}} u_j n_k dS + \int_{S_B} T_{ijk}u_j \frac{\partial}{\partial \boldsymbol{a}}[n_k dS] \\ &+ \int_{S_B} T_{ijk}\left(\frac{\partial u_j}{\partial \boldsymbol{a}}\right) n_k dS \end{aligned} \tag{12}$$

In these integrals, there are two new kernels which must be evaluated involving the derivatives of S_{ij} and T_{ijk}. These new integrals involve known quantities. The only unknown quantities appear in the last integral of (12). Combining the results of (9–12) in short form yields

$$\frac{\partial u_k}{\partial \boldsymbol{a}} = * \int_{S_B} T_{ijk} \left(\frac{\partial u_j}{\partial \boldsymbol{a}} \right) n_k dS + \quad \textit{known quantities} \tag{13}$$

This is an integral equation to be solved for each column of the Jacobian matrix corresponding to the given parameter a. The integral equation must be solved N times, but the kernel T_{ijk} and geometry is identical to the original boundary integral equation in each case. If the original matrix was inverted and saved, the solution for N sets of additional data requires only $O(N^3)$ additional operations. In summary, with the additional integrations and matrix solutions, a boundary integral solution plus Newton iteration requires approximately twice the effort of the boundary integral solution alone.

Once the matrix $\partial u_k / \partial \boldsymbol{a}$ has been found from the solutions of (13), the Jacobian matrix $\partial u_n / \partial \boldsymbol{a}$ is given by

$$\frac{\partial u_n}{\partial \boldsymbol{a}} = \frac{\partial}{\partial \boldsymbol{a}} (\boldsymbol{u} \cdot \boldsymbol{n}) = \frac{\partial u_i}{\partial \boldsymbol{a}} n_i + u_i \frac{\partial n_i}{\partial \boldsymbol{a}} \tag{14}$$

The inversion of the N'th order Jacobian matrix requires negligible effort compared to the inversion of the 3N'th order boundary integral matrix.

3. Numerical implementation. The numerical implementation of the boundary integral method is essentially the same for all the formulations discussed above. The required steps in the method are *(i)* geometry discretization, *(ii)* function discretization, *(iii)* equation discretization, *(iv)* integral evaluation and *(v)* solution of equations. When the geometry is known exactly, then an analytical specification may be used and there is no need for discretization. On the other hand, the geometry may not be known *a priori*, or it may prove inconvenient to employ the exact description. In these cases, the geometry must describe in an approximate fashion, typically using the same discretization as for the functions $\boldsymbol{u}$ and $\boldsymbol{f}$.

COORDINATE INTERVALS

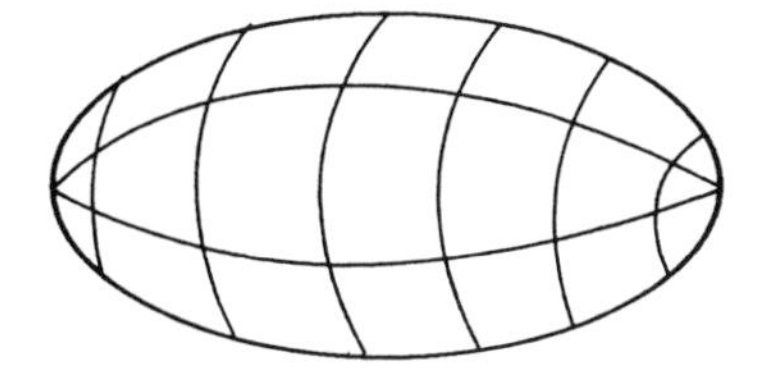

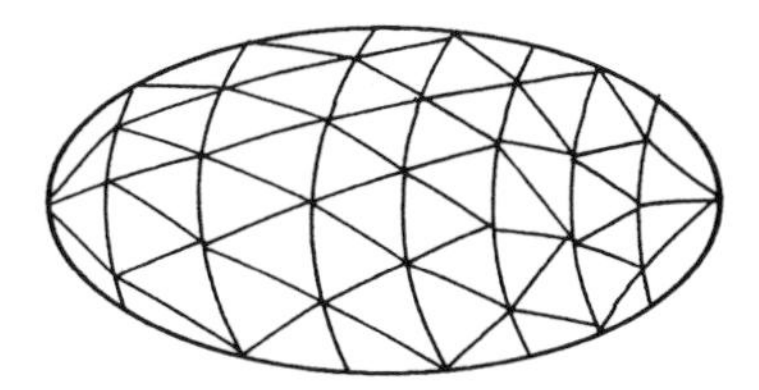

SPECTRAL ELEMENTS

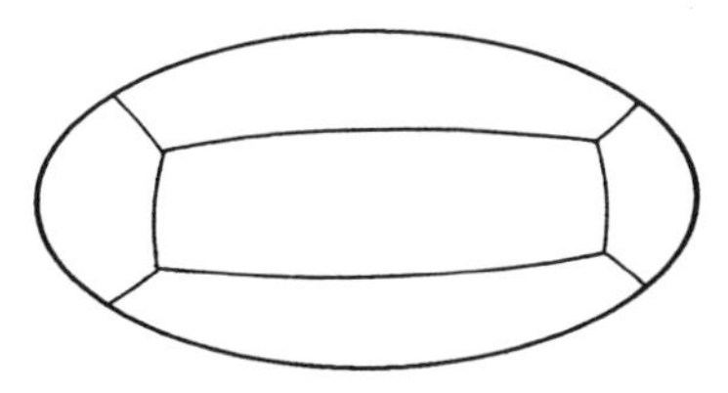

Figure 2 Discretization schemes for boundary surfaces in three dimensional boundary integral method.

Three possible discretization strategies for three dimensional flows are illustrated in Figure 2: coordinate specification, finite elements and spectral elements. The first method relies on the existence of a global coordinate system (e.g. spherical, cylindrical coordinates, etc.) in which the functions and geometry may be specified. Within this system, these variable may be defined in terms of spectral expansions (Fourier, spherical harmonics, etc) or at fixed coordinate intervals. Both approaches have been employed; their application to free surface flows will be described in section 4. These methods are the boundary integral analogs of spectral methods and finite difference techniques for partial differential equations. The other two discretizations – finite elements and spectral elements carry over directly from the respective techniques for differential equations. The use of finite elements is quite common in the boundary integral literature. Recent applications to Stokes flow include Chan et al [2] and Kennedy [7]. The use of spectral element techniques in boundary integrals is described in section 5.

Closely associated with the concept of function discretization is that of the equation discretization. In the case of partial differential equations, numerous strategies are used from pointwise collocation methods to integral Galerkin methods. Direct analogs may be employed in the boundary integral method, although collocation methods have proved to be the most popular choice, owing to the cost of integration. In a number of applications, especially with first kind equations, over-constrained least squares collocations methods have proved to be quite useful.

The formulation, discretization and solution of the boundary integral equations has received much attention in the literature, however, less attention has been paid to the evaluation of the singular integrals appearing in these equations. A number of methods have been proposed for dealing with the singularity. One may subtract off an analytical integral of the kernel over an approximate geometry, map the variables to a local polar coordinate system, or stretch the variables locally to eliminate the singularity. After one or more of these strategems, the remaining integral is typically evaluated using Gaussian quadratures.

While these approaches appear reasonable, there are two main objections. The first is that the integrations require the overwhelming majority of the computation time in most real three dimensional simulations – 90 to 95% of the cpu time is not uncommon. Given this dominance, additional effort in reducing this burden seems justified. The second objection is that most three dimensional simulations to date have been for simple geometries where the integrations are relatively easy. In fact it is not the singular integrals, but *nearly* singular integrals which often provide the most difficulty. Such integrals arise for example with intersecting surfaces or nearly touching tangent surfaces. Where quadrature counts of 4×4 or $t \times 6$ might have sufficed elsewhere, 30×30 or even 50×50 point Gaussian quadratures are now required for the same accuracy. The use of adaptive techniques (either automatic or manually tuned) can help control such problems, but they are not the final answer. If robust general purpose algorithms are to be developed, highly accurate, efficient quadrature techniques need to be developed.

On the subject of solution of equations, a few brief comments will suffice at this juncture. For equations of the first kind (6,7), no general iterative scheme has been discovered and direct Gaussian methods with $O(N^3)$ operations are employed. This expense is substantially ameliorated when the same equations need to be solved many times, because the inverse (or LU factors) may be saved yielding an $O(N^2)$ cost for additional solutions. Muldowney [12] describes some preliminary efforts at reducing the computational cost by applying a *partial* Gaussian reduction followed by an iteration on the reduced matrix. For equations of the second kind (completed double layer and free surface flows), iterative solutions (e.g. Jacobi type) may be used with $O(N^2)$ operation counts. While this represents a tremendous savings in principle, the actual savings in most applications to date has been negligible owing to the computational dominance of the numerical integrations. As larger problems are attacked in the future, and as more efficient quadratures are developed, the potential benefit of iterative solutions will be realized in practice.

This concludes our brief summary of the general principles of boundary integral methods. In the sections that follow, we examine the specific implementation of boundary integral methods in three distinct cases: a three dimensional coordinate based approach, a two dimensional spectral element method and an outline of a three dimensional spectral element method.

4. Three dimensional drops and bubbles. In this section, we discuss a coordinate based algorithm to model the deformation of drops and bubbles in linear extensional and shear flow. The velocity at the surface of the drop satisfies the boundary integral equation (8) discussed earlier. We shall be interested in both the unsteady evolution and steady equilibrium shapes of the drops. A full description of the algorithms and results is given by Schnepper [23]. In the development of the algorithm, tests were performed for rigid particles with 2:1 spheroid immersed in a uniform longitudinal flow as a standard test case. Nodal points were distributed over the surface at equal intervals in ϕ and according to a Gauss–Legendre formula in θ (at zeroes of $P_n(\cos\theta)$). In the first instance, an over-constrained least squares collocatation method was employed with constant basis functions on each element. Next a global spectral expansion in spherical harmonics was attempted with a Galerkin formulation. In each case, convergence was achieved to a relative accuracy of 3×10^{-4}, however, the computational cost was deemed unacceptable. In addition, it appeared that the accuracy and convergence rate would decrease rapidly as more deformed shapes were considered.

Based on these arguments, higher order local interpolant methods were developed. Again the nodal points were specified at (θ, ϕ) intervals with each nodal point in the center of its respective element (that is in the center of the individual θ and ϕ intervals). For a biquadratic method, the force distribution on any element was a Lagrangian interpolant based on nodal values on the 3×3 group of elements centered on that element. Similarly, for a biquartic method, the basis functions were 4th order polynomials in (θ, ϕ) based on a 5×5 element grouping. The interpolation elements were shifted as necessary for elements near the poles.

Number of area elements	$N_\theta = N_\phi$	Uniform Flow		Shear Flow	
		Biquadratic	Biquartic	Biquadratic	Biquartic
25	5	1.00040			
36	6	$1.0002\underline{6}$	$1.000\underline{20}$	$0.993\underline{63}$	0.99825
64	8	$1.0001\underline{0}$	$1.0001\underline{0}$	0.99805	$0.9996\underline{8}$
100	10	$1.000\underline{04}$	$1.0000\underline{1}$	0.99923	$0.9999\underline{3}$
Error $\sim N^{-\beta}$		$\beta = 3.5$	$\beta = 5.5$	$\beta = 4.1$	$\beta = 6.2$

Table 1

Comparison of biquadratic and biquartic interpolation for spherical coordinate based boundary integral method. Results for flow past a 2:1 spheroid with major axis along the z axis. Results for uniform flow are total force on particle divided by exact result. Results for shear flow are total moment divided by exact result. Underlined digits indicate uncertainty owing to numerical quadratures. Geometry is represented exactly.

A comparison of the biquadratic and biquartic methods was conducted for flow past a 2:1 spheroid in two test cases: a uniform axial flow and a transverse linear shear flow. Table 1 compares the results for the total force (uniform flow) and total torque (shear flow) on the particle. In each case, good accuracy was achieved with a convergence rate consistent with the order of the method. In these computations, quadrature errors were minimal and the geometry was calculated exactly yielding the best case results for the respective methods. Given the superiority of the biquartic method and the fact that the computational cost is only marginally higher than that of the biquadratic method, the higher order algorithm was used for all further computations.

	Biquartic - Exact Geometry			Biquartic - Approx Geometry		
N_θ	2:1	4:1	10:1	2:1	4:1	10.1
6	0.999849	1.005889	1.103551	1.004402		
8	0.999994	1.001971	1.078608	1.000256	1.139418	1.824155
12	1.000018	0.999855	1.045933	1.000027	1.043786	1.500083
16	1.000005	0.999757	1.026722	1.000015	1.013350	1.289642
20	1.000001	0.999888	1.015186	1.000014	1.004259	1.211241
24			1.008232		1.001202	1.177497

Table 2

Effect of geometry on convergence rate for biquartic boundary integral method. Results are for total force on a spheroid in uniform flow divided by exact value for three different aspect ratios. The exact columns use an analytical result for the geometry; the approximate columns use a spherical harmonic expansion for the geometry as described in the text.

The next series of tests documented the performance of the biquartic algorithm as more extreme geometries were considered and tested the effect of geometry discretization. The test case was an axisymmetric uniform flow past spheroids with aspect ratios of 2, 4 and 10. The total force was calculated using both the exact geometry and a global expansion for $r(\theta, \phi)$ in spherical harmonics. Owing to the expense of evaluating the spherical harmonics at a large number of quadrature points on the home element, a further local interpolation was introduced to reduce the computational expense. The results of these tests are shown in Table 2. Two important observations may be made. The first is that larger deformation requires much greater computational efforts for the same accuracy. With the exact geometry, a 6 point calculation for a 4:1 particle gives the same accuracy as a 24 point method for a 10:1 particle. The second observation is that for approximate geometry, the geometry discretization is the leading source of error by a substantial margin. The conclusion is that the resolution of the geometry approximation should exceed that of the function discretization.

Having completed the numerical tests, we turned our attention to the deformation of droplets in viscous flows. Two types of flows were considered, axisymmetric deformation in extensional flows and three dimensional deformation in linear shear flows. Time dependent calculations were performed to study a droplet evolving from an unperturbed spherical shape, while steady state Newton calculations were employed to determine equilibrium states. For time dependent calculations involving linear shear flows, the basic algorithm was modified to use a rotating coordinate system which kept the z axis aligned with the longest dimension of the deforming droplet. This assured an optimal geometry discretization and yielded the best accuracy. Let us define the capillary number $\Omega = 4\pi E\mu a/\gamma$ where E is the extension rate or shear rate, μ is the viscosity of the surrounding fluid, a is the radius of the initial sphere and γ is the surface tension. The viscosity of the droplet is $\lambda\mu$. The deformation of a droplet is characterized by a parameter $D = (L - B)/(L + B)$, where L and B are the largest and smallest dimensions of the deformed droplet.

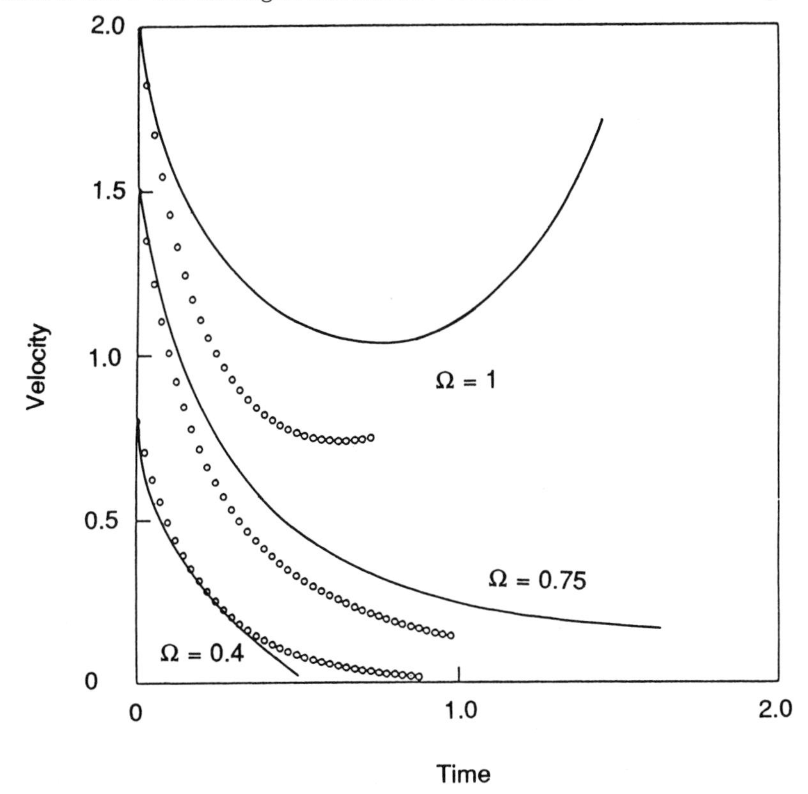

Figure 3 The deformation of a viscous droplet in a simple extensional flow. The endpoint velocity plotted versus time for three capillary numbers $\Omega = 4\pi E\mu a/\gamma$ and a viscosity ratio $\lambda = 1$. Open circles, Schnepper [23]; solid lines, Rallison and Acrivos [22].

The results of a typical unsteady calculation are shown in Figure 3 for the deformation of a drop in axisymmetric extensional flow at three different capillary numbers. The figure shows the velocity of the droplet endpoint as a function of time with open circles showing the present results and the solid lines those of Rallison and Acrivos [22]. For small capillary number $\Omega = 0.4$, the velocity quickly approaches zero. For large capillary number $\Omega = 1$, the velocity initially decreases retarded by interfacial tension; but then increases linearly as the endpoints are pulled along in the linear extensional flow. The intermediate value $\Omega = 0.75$ is quite near the critical capillary number. Our results have been confirmed by grid refinement and are accurate to within the plotted point size. The large apparent discrepancy of the earlier calculations is inconsequential as both calculations give similar values for the critical capillary number.

Flow	Ω	$\lambda = 1$	$\lambda = 0$
Extensional	0.4	0.119 0.122*	0.103 0.103*
	0.75	0.279 burst*	
	0.837758		0.245 0.246*
	1.0	burst	
Shear	2	0.177 0.178*	0.160
	3	0.268 0.278*	0.241

Table 3

Summary of results for deformation of droplets in linear flows: pure extension and simple shear, Schnepper [23]. Deformation D is defined as $(L + B)/(O - B)$ where L is major axis and B minor axis of droplet. Ω is capillary number $4\pi E \int a/\gamma$, λ is viscosity ratio (droplet/surrounding fluid). Where two value are given, the first is from unsteady calculation and second marked with $*$ is from steady state Newton calculation.

Table 3 summarizes the equilibrium shapes $D(\lambda, \Omega)$ for droplets in extensional and shear flows under varying conditions. Where two values are cited, the first is the result of the time dependent calculation and the second (marked with a *) comes from the steady state Newton calculation. In time dependent calculations, the velocity at the ends of the droplets becomes extremely small as equilibrium is approached. To minimize the computational expense, calculations were stopped when the endpoint velocity reaches a small cut-off value. This explains why the maximum deformation for the time dependent results is sometimes smaller than the steady state result. The results in Table 3 are broadly consistent with the numerical results of Rallison and Acrivos [22], Rallison [20] and with the asymptotic results of Taylor [25] and Chaffey and Brenner [1]. The results for inviscid droplets in linear shear flows are the first numerical results reported for such flows.

The biquartic boundary integral technique employed in these simulations yielded accurate and reliable results for the deformation of a single droplet in three dimensional flow fields. Despite this success, we are not satisfied with the performance of this algorithm. The deformation of the droplets was quite modest in most cases, and the computational effort for the three dimensional simulations was excessive. We believe that further algorithm development is warranted before proceeding to more complex flows. One possible avenue is explored in the following section.

5. Spectral element boundary integral formulation. In the spectral element approach, the geometry and the functions on the boundary are discretized in terms of high order polynomial approximations. In practice, we have often used polynomials of order $N = 8$ to 10, and have successfully employed orders as high as $N = 24$. We begin by discussing the method for two dimensional problems with a boundary integral formula analogous to (6).

Let the boundary surface of a 2D body be divided into a collection of surface elements S_α. On each surface element, define a parametric variable ξ on the interval $[-1, 1]$. This parametric variable might be normalized arc length, a local cartesian variable or any convenient quantity measuring position along the surface. Define collocation points along the surface $x(\xi_i)$ where the values of ξ_i are chosen as the zeroes of orthogonal polynomials, either Legendre or Chebyshev. With the points so chosen, the position of any point along the surface is given as a Lagrangian interpolant with respect to ξ; that is

$$\boldsymbol{x}(\xi) = \sum_{j=0}^{N} \boldsymbol{x}(\xi_j) h_j(\xi) \tag{15}$$

where h_j is the Lagrangian interpolating polynomial defined by $h_j(\xi_k) = \delta_{jk}$.

With these definitions, the shape of the boundary surface is defined as N'th order Chebyshev or Legendre polynomial in the variable ξ; here N is the number of collocation points per element. All quantities such as tangent vectors, normal vectors, curvature, etc may be calculated as derivatives of these functions. Because the polynomial representation is of high order, these derivative quantities do not suffer significant loss of accuracy.

Similarly, we define the physical variables $\boldsymbol{u}$ and $\boldsymbol{f}$ as Lagrangian interpolants based on their values at the same collocation points. This constitutes an *isoparametric* representation. When the order of the geometry discretization is higher or lower, we refer to a *superparametric* or *subparametric* representation respectively. In practice, it is most convenient to use the isoparametric form. Substituting the interpolants into the integral (6) leads to a linear system of algebraic equations

$$\mathbf{u} = \mathbf{A}\ \mathbf{f} + \mathbf{B}\ \mathbf{u} \tag{16}$$

The system matrices $\mathbf{A}$ and $\mathbf{B}$ are defined as integrals of the kernels and Lagrangian interpolants over the surface elements. Each matrix is of order $2N\ N_E$ where N is the order of the polynomials and N_E is the number of surface elements. The elemental submatrices $\mathbf{A}^{\alpha\beta}$ and $\mathbf{B}^{\alpha\beta}$ giving the velocity at points on element α owing to the integrals over element β are written

$$\begin{aligned} A(\alpha;\beta,j) &= \frac{1}{4\pi\mu} \int_{-1}^{1} \boldsymbol{S}(\boldsymbol{x}_\alpha, \boldsymbol{x}_\beta(\xi)) h_j(\xi)\omega d\xi \\ B(\alpha;\beta,j) &= \frac{-1}{4\pi\mu} \int_{-1}^{1} \boldsymbol{T}(\boldsymbol{x}_\alpha, \boldsymbol{x}_\beta(\xi)) \cdot \boldsymbol{n} h_j(\xi)\omega d\xi \end{aligned} \tag{17}$$

where ω is the area element $dS/d\xi$.

These integrals are evaluated by Gauss–Legendre quadratures with the aid of variable transformations. Briefly, there are four types of integrals which give slow convergence, each requiring a different coordinate stretch for efficient quadratures. These special cases are distinguish by the location of ξ^*, defined as the point on the element lying closest to the collocation point: *(i)* singular integrals with a collocation point on the element, *(ii)* point of closest approach ξ^* interior to the element, *(iii)* point ξ^* at end point with collocation point nearly on axis *(iv)* point ξ^* at end point with collocation point off axis. A comprehensive study of the transformations required in each case may be found in Muldowney [12].

N	Shear stress	Error
3	2.209 466 3	0.046 394 5
5	2.254 712 7	0.001 148 1
7	2.255 834 8	0.000 026 0
9	2.255 860 7	0.000 000 1
11	2.255 860 8	–

Table 4

Stokes flow over a wavy wall $y = 0.25 \sin x$ driven by translating rigid plate at $y = 1.25$. Shear stress at crest of wave as a function of N, the order of the spectral element representation. Domain is one period of wave with 10 elements around the boundary.

As a test problem for Stokes flow calculations with the spectral element boundary integral method, consider a simple shear flow over a sinusoidal wavy wall $y = .25 \sin x$. The domain consists of one period and its boundary is divided into ten elements. The shear stress at the crest of the wave as a function of N is shown in Table 4. These results show the exponential rate of convergence typical of spectral approximations: as N increases from 3 to 9, the error is reduced by a factor 4×10^5. This demonstrates that the spectral element formulation is a powerful technique capable of extraordinary accuracy with modest computational effort.

In the direct solution of partial differential equations, the rapid convergence of the spectral element approach is partially offset by the increased computational effort compared to low order methods. This increase is associated with the large bandwidth of the spectral method. In boundary integral methods, the system matrix is always full, hence there is no additional computational overhead. One may conclude that a spectral element formulation is the most efficient approach to high accuracy boundary integral computations. In two dimensional problems, the computational advantage is not decisive given the power of modern computers; for three dimensional problems it may prove essential.

In principle, the extension of the spectral element discretization to three dimensional flows is straightforward. To discretize a two dimensional boundary surface, suppose the surface is divided into curvilinear quadrilateral elements with parametric variables (ξ, η) each defined on the interval $[-1, 1]$. A point $\boldsymbol{x}(\xi, \eta)$ is defined in terms of the product interpolant

$$\boldsymbol{x}(\xi, \eta) = \sum_{j=0}^{N} \sum_{k=0}^{N} \boldsymbol{x}(\xi_j, \eta_k) h_j(\xi) h_k(\eta) \tag{18}$$

and the functions $\mathbf{u}$ and $\mathbf{f}$ are defined analogously.

The discretized linear system is given by (16) as before, while the elemental submatrices have the form

$$\begin{aligned} A(\alpha; \beta, j) &= \frac{1}{8\pi\mu} \int_{-1}^{1}\int_{-1}^{1} \boldsymbol{S}(\boldsymbol{x}_\alpha, \boldsymbol{x}_\beta(\xi, \eta)) h_j(\xi) h_k(\eta) \omega d\xi d\eta \\ B(\alpha; \beta, j) &= \frac{-1}{8\pi\mu} \int_{-1}^{1}\int_{-1}^{1} \boldsymbol{T}(\boldsymbol{x}_\alpha, \boldsymbol{x}_\beta(\xi, \eta)) \cdot \boldsymbol{n} h_j(\xi) h_k(\eta) \omega d\xi d\eta \end{aligned} \tag{19}$$

with $\omega = \partial(x, y)/\partial(\xi, \eta)$.

The only remaining difficulty is the evaluation of the 2D surface integrals in (19). The efficient evaluation of these integrals is crucial for the success of the method. One could apply a variety of coordinate stretches as for the one dimensional integrals, however this leads to rather large operation counts. In particular, one loses some of the advantage gained from the separability of the product interpolant. Muldowney explores the use of such coordinate transformations for two dimensional integration. Alternatively, one could divide each element into a number of subelements and employ analytical/numerical integration with modest quadratures to achieve operation counts comparable to those of low order discretizations. The accuracy would be higher, but one is still left with a very expensive integration algorithm. A possible alternative is outlined below.

The prototypical integral which must be evaluated in (19) is of the form

$$I = \int_{-1}^{1}\int_{-1}^{1} \frac{P_j(\xi)P_k(\eta)}{r}\, \omega d\xi d\eta \tag{20}$$

where $r^2 = (x - x_0)^2 + (y - y_0)^2 + (z - z_0)^2$ is a high order polynomial in (ξ, η).

In this expression, we find it more convenient to use the Legendre polynomials P_j in preference to the Lagrangian polynomials h_j. One may transform between the two with negligible computational expense. Suppose that the point $\boldsymbol{x}_0$ lies within the element at position (ξ_0, η_0). Expand $\boldsymbol{x}$ in a Taylor series about this point

$$x(\xi, \eta) = x_0 + x_\xi(\xi - \xi_0) + \xi_\eta(\eta - \eta_0) + \cdots \tag{21}$$

and express r^2 in the form

$$r^2 = a\widehat{\xi}^2 + 2b\widehat{\xi}\,\widehat{\eta} + c\widehat{\eta}^2 + O(\widehat{\xi}^3, \widehat{\eta}^3) \tag{22}$$

where $\widehat{\xi} = \xi - \xi_0, \widehat{\eta} = \eta - et_0$ and a, b and c are constants.

Define ρ^2 as the leading order approximation

$$\rho^2 = a\widehat{\eta}^2 + 2b\widehat{\xi}\,\widehat{\eta} + c\widehat{\eta}^2 \tag{23}$$

and expand $1/r$ as

$$\frac{1}{r} = \frac{1}{\rho} + O\left(\frac{\widehat{\xi}^3}{\rho^3}, \frac{\widehat{\eta}^3}{\rho^3}\right) \tag{24}$$

the leading term in the integral (20) is thus

$$I_{jk} = \int_{-1}^{1}\int_{-1}^{1} \frac{P_j(\xi)P_k(\eta)}{\rho}\, d\xi\, d\eta \tag{25}$$

These integrals I_{jk} may all be evaluated *analytically*. More importantly, with the analytical evaluation of I_{00}, I_{10}, I_{01} and a sequence of one dimensional integrals, all higher terms may be calculated by a simple *recursion formula*. Thus all integrals I_{jk} up to order N may be evaluated in order N^2 operations. The other terms involved in $\boldsymbol{S}$ and $\boldsymbol{T} \bullet \boldsymbol{n}$ yield to the same treatment, and a similar procedure may be employed for the non-singular integrals. Thus for quadrilateral planar elements all integrations could be performed analytically. For the general isoparametric mapped elements, the next term is $O(\widehat{\xi}^3/\rho^3, \widehat{\eta}^3/\rho^3)$. While this term in non-singular, it has singular first derivatives at (ξ_0, η_0) and hence convergence will be very slow even with Gaussian quadratures. One can repeat the above procedure and evaluate all of these integrals analytically. This process can be extended to arbitrary order yielding a remainder term which as smooth as one desires. The remainder could then be computed with a straight $N \times N$ Gauss product formula. The only difficulty is that the number of terms one deals with in the analytical integrations and recursion formulas becomes rather large leading to a significant computational effort and an administrative challenge. Nonetheless, the use of local expansions with analytical integration illustrates one possible approach for efficient computation of the singular and nearly singular integrals on mapped domains. The successful implementation of such a strategy would lead to a robust, accurate and extremely efficient implementation of the boundary integral method for three dimensional Stokes flows.

Acknowledgements. The authors would like to acknowledge the contributions of G. Muldowney, M. Cohn and J. Occhialini in the development of the spectral element of algorithms.

REFERENCES

[1] C.E. Chaffey and H. Brenner, *A second-order theory for deformation of drops*, J. Colloid Interfacial Sci., 24 (1967), pp. 258–269.

[2] C.Y. Chan, A.N. Beris and S.D. Advani, *Investigation of 3D hydrodynamic interactions around ellipsoidal particles using high order boundary element techniques*, AIChE National Meeting, (1990), Chicago, Illinois.

[3] M. Cohn, PhD dissertation, University of Illinois (1990).

[4] A.S. Geller, S.H. Lee and L.G. Leal, *The creeping motion of a spherical particle normal to a deformable interface*, J. Fluid Mech. 169 (1986), pp. 27–69.

[5] J.J.L. Higdon, *Stokes flow in arbitrary two-dimensional domains: shear flow over ridges and cavities*, J. Fluid Mech. 159 (1985), pp. 195–226.

[6] ——————, *Effect of pressure gradients on Stokes flows over cavities*, Phys. Fluids A 2, (1990), pp. 112–114.

[7] M. Kennedy, PhD dissertation, University of California, San Diego, (1991).

[8] S. Kim and S.J. Karrila, *Microhydrodynamics: Principles and Selected Applications*, Butterworth-Heinemann, Boston, 1991.

[9] R.E. Larson and J.J.L. Higdon, *Microscopic flow near the surface of two dimensional porous media. Part 1. Axial Flow*, J. Fluid Mech. 166 (1986), pp. 449–472.

[10] ——————, *Microscopic flow near the surface of two dimensional porous media. Part 2. Transverse flow*, J. Fluid Mech. 178 (1987), pp. 119–136.

[11] S.H. Lee and L.G. Leal, *The motion of a sphere in the presence of a deformable interface. II. A numerical study of the translation of a sphere normal to an interface*, J. Colloid Interfacial Sci. 87 (1982), pp. 81–106.

[12] G. Muldowney, PhD dissertation, University of Illinois, 1989.

[13] L. Newhouse and C. Pozrikidis, *The Rayleigh-Taylor instability of liquid layer resting on a plane wall*, J. Fluid Mech. 217 (1990), pp. 231–254.

[14] J.M. Occhialini, MS thesis, University of Illinois, 1989.

[15] H. Power and G. Miranda, *Second kind integral equation formulation of Stokes flow past a particle of arbitrary shape*, SIAM J. Appl. Math. 47 (1987), pp. 689–698.

[16] C. Pozrikidis, *The flow of a liquid film along a periodic wall*, J. Fluid Mech. 188 (1988), pp. 275–300.

[17] ——————, *The instability of moving viscous drops*, J. Fluid Mech. 210 (1990), pp. 1–21.

[18] ——————, *The deformation of a liquid drop moving normal to a plane solid wall*, J. Fluid Mech. 215 (1990), pp. 231–254.

[19] ——————, *Boundary integral and Singularity Methods for Linearized Viscous Flow*, Cambridge, UK, 1991.

[20] J.M. Rallison, *A numerical study of the deformation and burst of a viscous drop in general shear flows*, J. Fluid Mech. 109 (1981), pp. 465–482.

[21] *The deformation of small viscous drops and bubbles in shear flows*, Ann. Rev. Fluid. Mech. 16 (1984), pp. 45–66.

[22] J.M. Rallison and A. Acrivos, *A numerical study of the deformation and burst of a viscous drop in extensional flow*, J. Fluid Mech. 89 (1978), pp. 191–200.

[23] C.A. Schnepper, MS thesis, University of Illinois, 1989.

[24] H.A. Stone and L.G. Leal, *Relaxation and breakup of an initially extended drop in an otherwise quiescent fluid*, J. Fluid Mech. 198 (1989), pp. 399–427.

[25] G.I. Taylor, *The viscosity of a fluid containing small drops of another fluid*, Proc. Roy. Soc. A 138 (1932), 41–48.

[26] G.K. Youngren and A. Acrivos, *On the shape of a gas bubble in a viscous extensional flow*, J. Fluid Mech. 76 (1976), pp. 433–442.

LONG WAVE INSTABILITY OF VISCOUS LIQUID FREE SURFACE DUE TO ANOMALOUS MARANGONI EFFECT

V.V. PUKHNACHOV*

The phenomenon mentioned in the title of this paper is described in terms of solutions of the differential equation $u_t + \Delta^2 u + \Delta(u^2 - \beta u) = 0$, where Δ is the two-dimensional Laplacian, $\beta = \text{const}$. We study a qualitative behaviour of Cauchy problem solutions for this equation and give classification of their limiting regimes at $t \to \infty$. In particular, we formulate the sufficient condition of the solution collapse for a finite time.

1. Introduction. Let us consider a liquid layer limited by a solid bottom $x_3 = -h$ and by a free surface $x_3 = 0$. The bottom temperature θ_1 is assumed to be constant. A heat exchange between the liquid and the outer gas is governed by the Newton law, where the gas temperature $\theta_2 = \text{const} < \theta_1$. The gravity force acting along the x_3 axis has a constant acceleration $-g$. The liquid equilibrium which is possible in these conditions may be broken for two reasons. The first reason is a thermal expansion of a liquid. The basis of the corresponding stability theory was laid by Lord Rayleigh in 1916. The second mechanism of instability was discovered by J.R.I. Pearson in 1958. This is caused by the temperature dependence of the surface tension (the Marangoni effect). The Pearson instability condition is $Ma > Ma^*$, where $Ma = (\theta_1 - \theta_2)h\ |d\sigma/d\theta|/\mu\chi$ is the Marangoni number.

Here $\sigma(\theta)$ is the surface tension coefficient, μ is the dynamical coefficient of viscosity, χ is the thermodiffusion coefficient, Ma^* is the critical Marangoni number depending on the Biot number and the relation ℓ/h, where ℓ is the length of a disturbance wave. The detailed information concerning the stability of equilibrium of non-uniformly heated liquid in a horizontal layer is contained in the monograph [1].

2. Anomalous Marangoni effect. According to Gibbs thermodynamics of an interface, for pairs of pure liquids the dependence of the surface tension coefficient σ on temperature θ obeys the relation $d\sigma/d\theta = -s < 0$, where s is the specific surface entropy. For solutions the property of having a fixed sign of $d\sigma/d\theta$ is not necessary. R. Vochten and G. Pétré [2] revealed the anomalous $\sigma(\theta)$ dependence at the interface between out and n-heptanol solution in distilled water. This dependence is well described by the equation

$$\sigma = \sigma_0 + \phi(\theta - \theta_0)^2 \tag{2.1}$$

with appropriate positive constants σ_0, θ_0, ϕ. The non-monotonic character of the dependence $\sigma(\theta)$ has also been found in some liquid steels and alloys at the boundary with air (see [3,4] and the references given there).

*Lavrentyev Institute of Hydrodynamics, Siberian Division of the USSR Academy of Sciences, Novosibirsk 630090 USSR

Investigations of the minimum surface tension as a function of temperature on the convective flow structure have begun quite recently [5]. An interesting question has been posed in [5]. Let us suppose that in the equilibrium state of a liquid in a horizontal layer the temperature of a free surface equals to θ_0. Will this equilibrium be stable at any temperature drop $|\theta_1 - \theta_0|$, where θ_1 is the temperature of the bottom?

The answer to this question in terms of the linear theory will be positive since the corresponding Marangoni number which is proportional to $d\sigma(\theta_0)/d\theta$ will equal zero, in accordance with the formula (2.1). We should make a reservation beforehand that we do not know the answer if our problem is examined in an exact non-linear statement. However, some advancement is possible if we study a long-wave approximation in this problem.

3. Formulation of problem. We assume that the disturbance wave length ℓ is much more than the layer thickness h which, in its turn, is much more than the disturbance amplitude of free surface δ. In this case we come to the following equation for the non-dimensional deviation of a free boundary from a horizontal equilibrium state:

$$u_t + \Delta^2 u + \Delta(u^2 - \beta u) = 0. \tag{3.1}$$

Here $\Delta = \partial^2/\partial x_1^2 + \partial^2/\partial x_2^2, x_1$ and x_2 are the dimensionless coordinates in a layer plane, t is the suitable dimensionless time, and $\beta = \rho g h^2/\sigma_0$ is the Bond number (ρ is the liquid density). The case $\beta > 0$ corresponds to a normal direction of gravity.

The initial condition

$$u = u_0(x_1, x_2), \quad t = 0, \quad (x_1, x_2) \in \mathbb{R}^2 \tag{3.2}$$

joins to the equation (3.1). We suppose for simplicity that u_0 is the periodic function of both variables. Let us denote as Π the rectangle of periods of function u_0. We may consider that the greater side of Π equals 2π (one had to make an additional scaling transformation in an opposite case).

Thus, we study space-periodic solutions of Cauchy problem (3.1), (3.2). The following conservation law takes place for these solutions

$$\frac{d}{dt} \iint\limits_{\Pi} u \; dx_1 \quad dx_2 = 0. \tag{3.3}$$

Therefore, the mean value $\overline{u}$ of function $u(x_1, x_2, t)$ on the rectangle of periods does not depend on t. The case $\overline{u} = 0$ meets to the critical value θ_0 of temperature on the equilibrium free boundary. The common case may be reduced to the case $\overline{u} = 0$ using the substitution $v = u - \overline{u}$ and the replacement of parameter β in (3.1) on $\beta - 2\overline{u}$.

4. Historic survey. A one-dimensional variant of the equation (3.1) for a modelling of a surface deformation during a thermocapillary motioin was offered by

T. Funada [6]. Numerical simulation of 1–D periodic Cauchy problem was fulfilled in [7,8]. The second of these works contains also results of numerical analysis of 1–D Cauchy problem for decreasing initial data $u_0 = O(x_1^{-2})$ as $x_1 \to \infty$. Most interesting conclusion of both works is a the discovery of a tendency to a collapse of solution for some initial data. The proposal concerning the solutions of (3.1) destruction for a finite time was announced in [9,10]. The first of the cited works deals with the periodic Cauchy problem (3.1), (3.2) and the second one studies the initial boundary-value problem where the solution of (3.1) satisfies the condition $u = 0, \quad \Delta u = 0$ on the boundary of a finite domain $\Omega \subset \mathbf{R}^2$. There are other applications of (3.1) and this type of equations in the articles [11–13].

5. Stability of trivial solution. The equation (3.1) has a set of trivial solutions $u =$ const. In the light of comment at the end of Section 3, we can restrict our considerations by the case $u = 0$ without a loss of community.

And so, we consider a stability of solution $u = 0$ of the equation (3.1). The linearization of (3.1) on this solution leads to the equation $u_t + \Delta^2 u - \beta\Delta u = 0$. One may suppose that the rectangle of periods for function $u(x_1, x_2, t)$ at t fixed is

$$\Pi = \{x_1, x_2 : 0 < x_1 < 2\pi, \quad 0 < x_2 < 2\pi\, æ^{-1}\},$$

where æ≥ 1. Searching for a solution in the form

$$u = \mathrm{Re}\ \exp\{\lambda t + i(kx_1 + 1æ\, x_2)]$$

(here k and 1 are natural numbers), we obtain the following dispersion relation

$$\lambda = -k^4 - 2(kl\kappa)^2 - (1æ)^4 - \beta k^2 - \beta(1æ)^2. \tag{5.1}$$

In accordance with (5.1), the zero solution of linearized equation is stable for $\beta \geq -1$ and unstable for $\beta < -1$.

Before formulating the conditions of nonlinear stability for the trivial solution of (3.1), we must be sure that Cauchy problem (3.1), (3.2) has a solution for any $t > 0$ at least at a small (in an appropriate norm) initial data.

Let us denote as $H_0^s(\Pi)$ (s is a nonnegative integer) the subspace of Sobolev space formed by functions $w(x_1, x_2) \in H_{\mathrm{loc}}^s(\mathbf{R}^2)$ which are periodic on x_1 and x_2 with periods 2π and $2\pi æ^{-1}$, respectively, and which have the zero mean value on the domain Π. The next proposition is given without a proof.

Let $u_0 \in H_0^2(\Pi)$ and $\|u_0\|_{H^2} \leq \varepsilon$, where $\varepsilon = \varepsilon(\beta, æ)$ is a sufficient small positive number. If $\beta > -1$, Cauchy problem (3.1), (3.2) has a unique generalized solution $u(x_1, x_2, t) \in L^2(0, \infty; H_0^2(\Pi))$. Moreover, there exist constants $\gamma \in (0, 1 + \beta)$ and $C > 0$ independent of t such that the estimate $e^{\gamma t}\|u\|_{L^2} \leq C\varepsilon$ is true for any fixed $t > 0$.

This statement follows, in essence, from Theorem 2.2.5 of the monograph [14]. It means, in particular, the asymptotical stability of solution $u = 0$ of equation (3.1). One should note that the condition of smallness of $\|u_0\|_{H^2}$ is essential for the global existence of solution of problem (3.1), (3.2). We shall show in Section 8 that solutions having a "large" initial norm can be destroyed for a finite time.

As for the instability condition, Theorem 2.3.2 from [14] provides the possibility of concluding that a trivial solution of (3.1) is unstable in $L^2(\Pi)$ if $\beta < -1$. This result has a simple physical sense since the negative values of Bond number β correspond to a situation of "liquid layer on the ceiling".

6. Lyapunov functional. The following identity plays the key role in our further considerations:

$$\frac{dS(u)}{dt} = \iint_{\Pi} |\nabla[\Delta u + (u - \beta/2)^2]|^2 \, dx_1 \, dx_2. \tag{6.1}$$

Here $S(u)$ is the functional defined by the equality

$$S(u) = \iint_{\Pi} \left[\frac{1}{3} \left(u - \frac{\beta}{2} \right)^3 - \frac{1}{2} |\nabla u|^2 \right] dx_1 \, dx_2, \tag{6.2}$$

u is an arbitrary solution of (3.1) which satisfies the above-mentioned condition of periodicity.

The proof of the identity (6.1) consists in a direct calculation of dS/dt using equation (3.1) for substitution u_t, integration by parts and periodicity condition. The relation (6.1) is the immediate generalization of the identity put by V.I. Yudovich [9] for the 1–D analogue of equation (3.1) in the case $\beta = 0$.

As a consequence of (6.1), the function $S(u)$ does not decrease with the growth of t on any solution of equation (3.1). This is the property which gives the reason for us to call $S(u)$ the Lyapunov functional for (3.1). It is easy to see that the functional $S(u)$ increases strictly with t if the solution $u(x_1, x_2, t)$ is not a stationary solution of (3.1).

Let us consider the functional $S(u)$ on a smooth function $u(x_1, x_2, t)$ which is not necessarily a solution of Cauchy problem (3.1), (3.2) but which satisfies the same periodicity condition and which has the zero mean value on Π. We denote as δu a variation of u, moreover δu is assumed to have analogous properties as u. On the basis of (6.2), the first variation of $S(u)$ is

$$\delta S = \iint_{\Pi} (\Delta u + u^2 - \beta u) \delta u \, dx_1 \, dx_2 \tag{6.3}$$

(we took into account that $\delta \overline{u} = 0$). The last equality means that $\mathrm{grad}_{L^2} S(u) = \Delta u + u^2 - \beta u$. Hence, the equation (3.1) may be written in the form of $u_t = -\Delta \left[\mathrm{grad}_{L^2} S(u)\right]$.

We notice now that in the scale of Sobolev spaces $H_0^s(\Pi)$ a symmetric and positively defined operator $-\Delta$ has the bounded inverse $-\Delta^{-1} : H_0^s \to H_0^{s+2}(\Pi)$. It allows the equality (6.3) to be rewritten as

$$\delta S = \iint_{\Pi} [-\Delta(\Delta u + u^2 - \beta u)] \Delta^{-1} \, \delta u \, dx_1 \, dx_2. \tag{6.4}$$

If at present $\Delta^{-1}\delta u \in H_0^2(\Pi)$ and $u \in H_0^2(\Pi)$, then $-\Delta(\Delta u + u^2 - \beta u) \in H^{-2}(\Pi)$ and $\text{grad}_{H^{-2}} S(u) = -\Delta(\Delta u+u^2-\beta u)$ owing to (6.4) (The definition and properties of spaces $H^{-s}, s > 0$, see, for instance, in [15]). We infer from here that equation (3.1) has an equivalent form,

$$u_t = \text{grad}_{H^{-2}} S(u), \tag{6.5}$$

where the functional $S(u)$ is defined by (6.2). The gradient form (6.5) of equation (3.1) is useful for the investigation of qualitative properties of solutions of Cauchy problem (3.1), (3.2).

7. Stationary solutions and their stability. Due to Liouville's theorem, each stationary solution of (3.1) defined and bounded in $\mathbf{R}^2$ simultaneously is the solution of the equation $\Delta u_s + u_s^2 - \beta u_s - C = 0$, where $C = \text{const}$. For solutions $u_s(x_1, x_2)$ from the class $H_0^2(\Pi)$ we have

$$\Delta u_s + u_s^2 - \beta u_s - \overline{u_s^2} = 0, \tag{7.1}$$

where $\overline{u_s^2}$ is the mean value of the function u_s^2 in the domain Π.

We do not dispose of the full description of solutions $u_s \in H_0^2(\Pi)$ of equation (7.1). However, it is not difficult to find the "small" solutions of (7.1) with the help of the bifurcation theory methods (see, for example, [16]). Omitting details, we tell the final result. Each value $\beta_{k1} = -k^2 - 1^2$ æ^2 (k and 1 are non-negative integer numbers and $k1 > 0$) is a bifurcation point for the equation (7.1).

A special place occupies solutions of equation (7.1) which depends on one variable (say, x_1) only. In this case our equation is integrated in elliptic functions; corresponding periodic solutions are the well-known cnoidal waves [17]. At last, besides the small double-periodic solutions, equation (7.1) has a set of solutions with a hexagonal symmetry. If the side of the hexagon is $4\pi/3$ then values $\beta_k = -k^2$ are point of bifurcation for equation (7.1) (which is considered now in the class of functions possessing an appropriate symmetry).

Another interesting set of stationary solutions of (3.1) consists of solutions which are submitted to some conditions of the decrease when one or both variables tend to infinity. Such solutions satisfy the equation

$$\Delta u_s + u_s^2 - \beta u_s = 0 \tag{7.2}$$

for $(x_1, x_2) \in \mathbf{R}^2$. Let us suppose that $u_s = f(x_1)$ and $f \to 0$ when $x_1 \to \pm\infty$. The corresponding boundary-value problem for an ordinary differential equation concerning the function f has a solution if $\beta > 0$ only. This solution is a classical solution of Korteweg and de Vries,

$$f = \frac{3\beta}{2\ ch^2(\beta^{1/2}x_1/2)}\ . \tag{7.3}$$

It is interesting to find an asymptotical analogue of the solution (7.3), $u_s = g(r)$, where $r = (x_1^2 + x_2^2)^{1/2}$. The function g is the solution of the following boundary value problem:

$$\frac{d^2g}{dr^2} + \frac{1}{r}\frac{dg}{dr} + g^2 - \beta g = 0, \quad r > 0, \tag{7.4}$$

$$|g| < \infty \quad \text{at} \quad r \to 0, \quad g \to 0 \quad \text{at} \quad r \to \infty. \tag{7.5}$$

It turned out that for any $\beta > 0$ the problem (7.4), (7.5) has at least one non-negative solution. We give here the sketch of the proof of this proposition only.

Since main difficulties of our problem are connected with the non-compactness of the domain $r > 0$, we study at first the regularized problem where a nonlinear term in (7.4) is replaced by ηg^2. Here $\eta(r) \in C^\infty[0, \infty)$ is the cutting function so that $0 \le \eta \le 1$, if $r \ge 0$ and $\eta = 1$ if $r \in [0, N]$, $\eta = 0$ if $r \ge 2N$. The solvability proof for the regularized problem is based on the ideas of the work [18]. We search its solution $g_N(r)$ as $\lambda_N^{-1} y_N$, where y_N is a solution of the variational problem

$$I(y_N) \equiv \int_0^\infty r\left[\left(\frac{dy_N}{dr}\right)^2 + \beta y_N^2\right] dr \to \min$$

at the additional condition

$$J(y_N) \equiv \int_0^\infty r\eta\, y_N^3\, dr = C = \text{const} > 0,$$

λ_N is a normalizing coefficient of Lagrange. We proof the existence of a weak limit of a minimizing sequence for the functional I, $y_N^{(n)} \xrightarrow{w} y_N$ when $n \to \infty$. Simultaneously we get the estimate $\lambda_N \ge \Lambda > 0$ ($\Lambda(C, \beta)$ does not depend on N), when $N \to \infty$, this estimate will guarantee a non-triviality of the solution of (7.4), (7.5). In order to realize a limiting transition as $N \to \infty$, we use the integral representation of g_N,

$$g_N(r) = \int_0^\infty K(r, s)\, \eta(s)\, s\, g_N^2(s)\, ds,$$

where $K(r, s)$ is the Green function of the Bessel operator $-d(rd/dr)/dr + \beta r$ with boundary conditions (7.5). On the basis of Hölder inequality, a uniform boundedness of $J(g_N)$ and the explicit expression of $K(r, s)$ in terms of Bessel functions, we obtain in two stages the estimate $g_N = O(r^{-4/3})$ when $r \to \infty$ uniformly concerning N. This estimate allows the passage to the limit in the sequence g_N when $N \to \infty$. The non-negativeness of g_N (and the limiting function $g(r)$) follows from the above integral representation if we take into account that $K(r, s) > 0$ when $r > 0, s > 0$. In fact, it is not difficult to get the inequality $g > 0$ at any finite $r > 0$ so that the obtained solution of problem (7.4), (7.5) is a non-trivial one. A posteriori we prove that the function $g(r)$ has an exponential decay when $r \to \infty$, similarly to the classical soliton.

Thus we see that the equation (3.1) owns by a reach supply of stationary solutions. On the other hand, it has no solutions in the form of travelling waves which are defined and bounded at all values of its arguments and which satisfy some natural conditions at infinity. Actually, let $u = q(x_1 - ct)$ be the solution of this type where $c = \text{const} \neq 0$. The ordinary differential equation for $q(\zeta)$ may be integrated one time; the result will be

$$\frac{d^3 q}{d\zeta^3} + \frac{d}{d\zeta}(q^2 - \beta q) - cq = 0$$

(if $c \neq 0$, the constant of integration can be chosen by zero due to a suitable change of parameter β). Assuming that $q(\zeta)$ decreases sufficiently quickly together with its derivatives when $\zeta \to \pm\infty$ we multiply the last equation on q and integrate the obtained equality. It gives

$$c \int_{-\infty}^{\infty} q^2 \, d\zeta = 0$$

and consequently $q = 0$.

Let us discuss now briefly the question of stability of stationary solutions. We restrict our considerations by the simplest case when the integrated solution $u_s \in H_0^2(\Pi)$ and its disturbances belong to $H_0^2(\Pi)$ also for any fixed $t \geq 0$.

From the results of section 6 it follows that each stationary solution u_s of equation (3.1) is the extremal point for the functional $S(u)$ which is considered on functions $u \in H_0^s(\Pi)$, $s \geq 1$ (the variable t, which may enter a number of arguments of the function u, plays a role of parameter in this part of Section 7). This circumstance, together with the gradient form (6.5) of equation (3.1) gives us a chance to apply an approach based on the Lagrange theorem about stability of an equilibrium state in classical mechanics. As is known, the analysis of the second variation $\delta^2 S$ of the functional $S(u)$ at the point u_s lies in the ground of this approach. An insignificant difference in the termsadopted in our case and a classical one is due to the fact that S plays a role in the entropy of a dynamical system associated with the equation (3.1) as the corresponding functional in the classical mechanics is the potential energy of a system.

According to (6.3), the expression for $\delta^2 S(u_s)$ has the form

$$\delta^2 S(u_s) = \iint_{\Pi} [-|\nabla \delta u|^2 + (2u_s - \beta)\,(\delta u)^2] \, dx_1 \, dx_2. \tag{7.6}$$

It is clear from (7.6) that the function S has no local minima. The definition (6.2) of $S(u)$ shows that this functional is not below or upper bounded. Typically, critical points of S are saddle points. However, if the following unilateral inequality

$$2u_s < 1 + \beta \tag{7.7}$$

is fulfilled, the quadratic form $\delta^2 S(u_s)$ is a positively definite one. This statement follows from (7.6) if it is taken into account that the constant is Steklov – Poincaré inequality for the rectangle Π equals a unity, and this value is an exact one.

If inequality (7.7) is not fulfilled at some points of the domain Π, the form $\delta^2 S(u_s)$ is indefinite, and we should expect the instability of the stationary solution u_s. Just like this solution takes place in the case $\beta < -1$, when the equation (7.1) has wittingly solution $u_s \in H_0^2(\Pi)$. (Really, for such solutions $\overline{u_s} = 0$ and $\max u_s \geq 0$, $(x_1, x_2) \in \Pi$). And though the inverse Lagrange theorem is not proved in our case, its assertion seems to be highly likely.

As for the direct Lagrange theorem, it is true for the dynamical system under our consideration, and we are able to confirm the stability of stationary solutions subordinated to condition (7.7). However, the check-up of this condition for $\beta > -1$ is a rather difficult procedure which can be made, probably, by numerical methods for a concrete solution u_s. In any case, this conditions is very hard, and at this moment we have no stationary solutions (except for $u_s = 0$) which satisfy (7.7).

8. Collapsing solutions. In the light of the results of section 6, 7 we may expect the following alternative at the evolution of the solution of Cauchy problem (3.1), (3.2): or $u \to u_s$ when $t \to \infty$, where u_s is some stationary solution of (3.1), or its solution is destroyed for a finite or infinite time. We formulate below some sufficient conditions for the realization of the second possibility.

Our examination is founded on works [19, 20]. Moreover, the demanded result will be obtained by the verification of conditions of Theorem 2.2 from [20]. The equations

$$Pu_t = -Au + Bu + F(u,t) \tag{8.1}$$

are considered in this article. Here P, A and B are linear operators, which are defined on the dense set $\mathcal{D} \subset \mathcal{H}$, where $\mathcal{H}$ is the Hilbert space, and P is the symmetric positive operator, A is the symmetric non-negative one, the operator B is subordinated, in a certain sense, to operators $A^{1/2}$ and $P^{1/2}$. The nonlinear term $F(t,u)$ is (at t fixed) Fréchet differential of some functional $G(t,u)$, which besides depends smoothly on t. (We keep the system of names of the work [20].)

In our case $A = -\Delta$, $P = A^{-1}$, $B = -\beta I$, $F = u^2$; we can take the set of smooth functions, which are periodic on x_1 and x_2 with periods 2π and 2π æ$^{-1}$ accordingly and which have zero mean value in Π, as $\mathcal{D}$. As for $\mathcal{H}$, one is most simple to choose $\mathcal{H} = H_0^0(\Pi)$ (see Section 5). by this we had to take care so as a number of terms arising in the proof of Theorem 2.2 [20], did not lose a sense (for example, $\|A^{1/2}u\|$). But we are able to achieve this choosing sufficiently smooth initial data $u_0 \in H_0^s(\Pi)$. In fact, until the collapse coming, the function u will be limited, and the mechanism of the smoothness rise for the solution of the parabolic equations (3.1) acts.

Principal conditions of Theorem 2.2 from [20] are $(F(t,u),u) \geq 2(1+\alpha_1)G(t.u)$, where $\alpha_1 > 0$, on the investigated solution u of the equation (8.10, and the inverse inequality $\nu G(t,u) \geq (F(t,u),u)$ with some $\nu > 0$ (here (u,v) denotes the scalar product $u, v \in \mathcal{H}$). In our case F does not depend on t explicitly;

$$G(u) = \frac{1}{3} \iint\limits_{\Pi} u^3(x_1, x_2, t)\, dx_1\, dx_2,$$

so that both above-mentioned inequalities transform into equalities if we choose $\alpha_1 = 1/2, \quad \nu = 3$. The expression $\|P^{1/2}u\|$, entering the formulation of Theorem 2.2, is now $L^2(\Pi)$ - norm of the function $(-\Delta)^{-1/2}u$; if $u \in H_0^s$ for fixed $t \geq 0$ at $s \geq 0$, the last function is defined and belongs to $H_0^{s+1}(\Pi)$.

PROPOSITION. *Let initial function $u_0 \in H_0^2(\Pi)$ in the condition (3.2) satisfy the inequality*

$$\iint\limits_{\Pi} \left(\frac{u_0^3}{3} - \frac{1}{2} |\nabla u_0|^2 \right) dx_1\, dx_2 > \tag{8.2}$$

$$> \frac{6}{5}\,(1+\beta^2) \iint\limits_{\Pi} [(-\Delta)A^{-1/2}u_0]^2\, dx_1\, dx_2.$$

There exists such $t^* > 0$ that for solution u of Cauchy problem (3.1), (3.2) we have

$$\|(-\Delta)^{-1/2}u\|_{L^2} \to \infty \quad \text{when} \quad t \to t^* - 0. \tag{3.1}$$

This proposition is an immediate corollary of Theorem 2.2 and preceding Theorem 2.1 from [20]. The inequality (8.2) is equivalent to the key condition (2.6) in the formulation of Theorem 2.1 if we fix the following values of constants entering there: $\alpha = 1/5, \quad \varepsilon = 19/269$. Cited theorems give also the upper estimate of the life time for the collapsing solutions. However, this estimate is rather cumbersome, and we do not adduce it here. We mark also that for $\beta = 0$ the collapse condition (8.2) may be relaxed; exactly, we can replace its right side by zero in this case.

It should be emphasized that the inequality (8.2) cannot be fulfilled for "small" data u_0. This inequality and its more weak variant at $\beta = 0$ are not valid also for functions $u_0 = C \cos\, kx_1 \cos 1æx_2$ (k and 1 are integers and $k^2 + 1^2 > 0$) at any value $C =$ const. We indicate a simple example of an initial function of Cauchy problem (3.1), (3.2) with $\beta = 0$, when the collapse takes place: $u_0 = a_1 \cos\, x_1 + a_2 \cos\, 2x_1$, where constants a_1 and a_2 submit to inequalities $|a_1| > 2, a_1^2 - (a_1^4 - 16)^{1/2} < 2a_2 < a_1^2 + (a_1^4 - 16)^{1/2}$.

The stated considerations allow to come near to the answer to the question put in the article [5] (see Section 2). We can confirm with a considerable probability that the discussed equilibrium state of a liquid layer is unstable with respect to long waves of a finite amplitude.

9. Open questions. a) As follows from the discussion in Section 7, one-periodic and double-periodic of the equation (7.1), as a rule, are unstable. It is interesting to study the stability of soliton-type solutions. Some understanding may be achieved here during an analysis of the spectral problem for the operator

$\Delta + 2f - \beta$ in $\mathbf{R}^2$, where the function f is defined by (7.3) (see, for instance, [21]), and an analogous problem for the operator associated with axisymmetrical soliton.

b) If $\beta = 0$ (a liquid is under weightlessness) the equation (3.1) has self-similar solutions in the form $u = t^{-1/2}\varphi(rt^{-1/4})$, where $r = (x_1^2 + x_2^2)^{1/2}$. These solutions satisfy the mass conservation law which takes place for the equation (3.1). (It is interesting to note that self-similar solutions of the 1–D analogue of (3.1) are not compatible with the mass conservation law). Yet, from the point of view of a singularity structure near a catastrophe moment t^*, other self-similar solutions are of interest: $u = (t^* - t)^{-1/2}\psi[r(t^* - t)^{-1/4}]$. The ordinary differential equation for function $\psi(\xi)$ is arranged highly hardly: if a corresponding boundary value problem is solvable in the class $\xi\psi(\xi) \in L^1(0, \infty)$, then its solutions are insulated. The question of the solvability for mentioned problem remains to be open.

c) The opportunity of blowing-up for solutions of the problem (3.1), (3.2) needs a correct physical interpretation. We believe that the reason of this phenomenon is a model character of the equation (3.1). We remind the hierarchy of scales at obtaining this equation: $\delta \ll h \ll \ell$ (see notations in Section 3). If we assume now that quantities δ and h have the same order acting in the spirit of the work [22], we lead to the next equation concerning function $u(x_1, x_2, t)$:

$$u_t + \nabla \cdot [(1 + \varepsilon u)^3 \nabla(\Delta u - u) + (1 + \varepsilon u)^2 (1 - \alpha\varepsilon u)^{-3} \nabla u^2] = 0. \tag{9.1}$$

Here $\varepsilon = \delta/h$, $\alpha \in (0.1)$ is some constant. The last equation turns into (3.1) when $\varepsilon \to 0$. If the parameter ε is small, we may hope to a similar behaviour of solutions for Cauchy problem (3.1), (3.2) and (9.1), (3.2) until the solution of the first problem is bounded. As for the problem (9.1), (3.2), there are definite grounds to believe that its solutions, which are limited from below by $-1/\varepsilon$ (the physical origin of (9.1) demands the non-negativeness of the function $1 + \varepsilon u$) will be limited from above also. It is an intriguing question: to investigate Cauchy problem (9.1), (3.2) (using analytical and numerical methods) and to find, in particular, the field of applicability for the shortening equation (3.1).

Acknowledgements. The author is extremely thankful to Professor V.I. Yudovich for numerous stimulating discussions and to Professor P.I. Plotnikov for useful consultations.

REFERENCES

[1] Joseph, D.D., *Stability of fluid motions*, Springer–Verlag. Berlin, Heidelberg, New York, 1976.

[2] Vochten, R., Pétré G., *Study of the heat of reversible adsorption at the air-solution interface. II. J. Colloid*, Interface Sci., 1973, Vol. 42.

[3] Villers, D., Platten, J.K., *Rayleigh–Bénard instability in system presenting a minimum surface tension*, In: Proc. 5th Europ. Symp. Material Sciences under Microgravity. Schloss–Elmau, 1984.

[4] Tretyakova, E.E., Baum, B.A., Tyagunov, G.V., Klemenkov, E.A., Talapina, M.V., Thepelev, V.S., Dunaev, V.G., *The temperature influence on the surface tension of liquid steels and alloys*, Izvestiia VUZov, Chyornaia metallurgiia, 1986, No. 6 (in Russian).

[5] LEGROS, J.C., LIMBOURG–FONTAINE, M.C., PÉTRE, G., *Influence of a surface tension minimum as a function of temperature on the Marangoni convection*, Acta Astronautica, 1984, Vol. 11, No. 2.

[6] FUNADA, T., *Nonlinear diffusion equation for interfacial waves generating by the Marangoni effect*, Note of Res. Inst. Math. Sci., Kyoto Univ., 1984, No. 510 (in Japanese).

[7] FUNADA, T., KOTANI, M., *A numerical study of nonlinear diffusion equation governing surface deformation in the Marangoni convection*, J. Phys. Soc. of Japan, 1986, Vol. 55, No. 11.

[8] PUKHNACHOV, V.V., *Manifestation of an anomalous thermocapillary effect in a thin liquid layer*, In: Gidrodinamika i teploobmen techenii zhidkosti so svobodnoi poverkhnost'yu. Novosibirsk, 1985 (in Russian).

[9] PROTOPOPOV, B.E., PUKHNACHOV, V.V., YUDOVICH, V.I., *An anomalous thermocapillary effect and its displays in weak force fields*, In: 4 Vsesoyuznyi seminar po gidromekhanike i teplomassoobmenu v nevesomosti. Tezisy dokladov, Novosibirsk 1987 (in Russian).

[10] KALANTAROV, V.K., *On a global behaviour of solutions of some fourth-order nonlinear equations*, In: Kraevye zadachi matematicheskoi fiziki i smezhnye voprosy teorii funktsii, Vol. 19, Nauka, Leningrad, 1987 (in Russian).

[11] KURAMOTO, Y., *Phase dynamics of weakly unstable periodic structures*, Progress of Theoretical Physics, 1984, Vol. 71, No. 6.

[12] KAWAHARA, T., TOH, S., *An approximate equation for nonlinear cross-field instability*, Phys. Letters, 1985, Vol. 113A, No. 1.

[13] SIVASHINSKY, G.I., *On cellular instability in the solidification of a dilute binary alloy*, Physica D., 1983, Vol. 8.

[14] YUDOVICH, V.I., *The method of linearization in the hydrodynamic stability theory*, Rostov na Donu, 1984 (in Russian).

[15] LIONS, J.–L., MAGENES, E., *Problèmes aux limities non homogènes et applications*, Dunod, Paris, 1968.

[16] IOOSS, G., JOSEPH, D.D., *Elementary stability and bifurcation theory*, Springer–Verlag. New York, Heidelberg, Berlin, 1980.

[17] WITHAM, G.B., *Linear and nonlinear waves*, John Viley & Sons. New York – London – Sydney – Toronto, 1974.

[18] POKHOZHAEV, S.I., *About eigenfunctions of equation* $\Delta u + \lambda f(u) = 0$, Soviet Math. Doklady, 1965, Vol. 165, No. 1 (in Russian).

[19] LEVINE, H.A., *Some nonexistence and instability theorems for formally parabolic equations of the form* $Pu_t = -Au + F(u)$, Arch. Rational Mech. Anal., 1973, Vol. 51.

[20] KALANTAROV, V.K., LADYZHENSKAYA, O.A., *On the rise of collapses for quasilinear parabolic and hyperbolic equations*, In: Krayevye zadachi matematicheskoi fiziki i smezhnye voprosy teorii funktsii, Vol. 10, Nauka, Leningrad, 1977 (in Russian).

[21] TITCHMARCH, E.C., *Eigenfunction expansions associated with second-order differential equations*, Calderon Press, Oxford, 1946.

[22] KOPBOSYNOV, B.K., PUKHNACHOV, V.V., *Thermocapillary motion in thin liquid layer*, In Gidromekhanika i protsessy perenosa v nevesomosti. Sverdlovsk, 1983 (in Russian).

LONG-WAVE INTERFACIAL INSTABILITIES IN VISCOUS LIQUID FILMS

MARC K. SMITH*

Abstract. One well-known instability that can occur in a viscous liquid film produces a wave-like deformation of the free surface over a wavelength that is much larger than the thickness of the film. We shall examine this motion and explain the dynamics of the instability mechanism by carefully interpreting the mathematical approximations used to solve the stability problem.

Several different geometries will be considered. The most fundamental is the isothermal liquid film on an inclined plane. Here, we explain how the small effects of liquid inertia lead to the unstable interfacial motion. Next, we consider a concentric arrangement of two fluids flowing in a vertical pipe. Here, the unstable behavior of the interface between the two fluids is modified by a large lubrication pressure that appears because of the rigid pipe wall. Thermal effects are explored by cooling an inclined liquid layer from below. In this case, both buoyancy forces and direct liquid expansion have important effects. Finally, a horizontal layer driven by thermocapillarity is considered to demonstrate the effect surface-tension gradients have on the instability.

1. Introduction. Thin liquid films are seen in many different engineering technologies. In some coating processes, e.g., the production of photographic film, one or more thin liquid layers are deposited onto a solid substrate. Some heat exchangers use a liquid film to cool a solid surface, and in others a liquid condensate film forms on the cold surface used to cool a hot gas. Thin films of low-viscosity fluid are also used to lubricate the flow of a high-viscosity liquid in a pipe in order to lower the pumping power needed to transport the liquid. An oil pipeline is one very important use of this technology.

In all of these processes, a uniform stable film is desired. Significant nonuniformities in the film thickness could produce an unacceptable product in the case of coating processes, a lowering of the efficiency of the heat transfer devices, or an increase in the pumping power in the fluid transport systems. In addition, a rupture of the film could cause a failure of the entire device or process.

To maintain a uniform film thickness, the design of the process or device must take into account any potential for an interfacial instability. In those applications where the lateral dimensions of the liquid film are much larger than the mean film thickness, it is possible for a long-wave instability to form in the system. This kind of instability has long fascinated both engineers and mathematicians because it is prevalent in so many different physical situations and because its mathematical description is amenable to a simple, yet robust perturbation technique.

Experiments have shown that the long-wave interfacial instability of the film first appears as a regular, two-dimensional wave with a small amplitude and a wavelength much larger than the film thickness. This suggests the use of linear hydrodynamic stability theory to describe the onset of the motion. In addition, the large wavelength suggests the use of a regular perturbation method for small wavenumbers. The result of the analysis is a simple and concise representation for the critical point of the film.

* The George W. Woodruff School of Mechanical Engineering, Georgia Institute of Technology, Atlanta, GA 30332-0405.

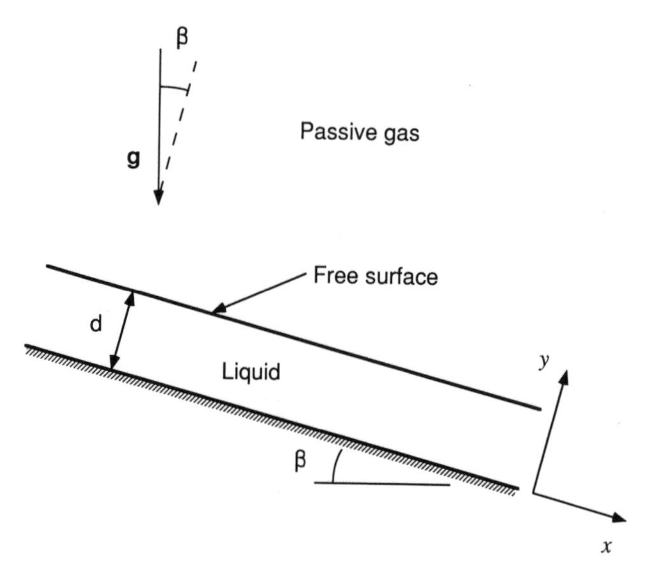

FIG. 1. *The geometry of an isothermal liquid layer on a rigid inclined plane.*

This method has been used by many researchers in the past thirty years or so to study the long-wave instability of liquid films in many different physical situations.

This unstable film flow is an example of a viscous flow with a deformable free surface. The complete description of the motion of the unstable interface as it evolves in time is a difficult nonlinear problem. It can be somewhat simplified for the case of long waves on thin films by using the long-wave nature of the motion to derive a nonlinear evolution equation for the interface as shown by Benney (1966) and Atherton & Homsy (1976). However, the goal of the present paper is to describe the dynamics of the long-wave instability in thin liquid films. We shall show that the mechanism of this instability is intimately connected to the interaction between the flow in the bulk liquid and the deformable interface. Fortunately, linear stability theory allows us to isolate this interaction in a relatively clean manner by reducing the free boundary problem to a problem with fixed boundaries. The cost we pay is a more complicated boundary condition on the formerly free upper surface.

With the many dozens of papers published on this instability, it seems as if almost any effect you like can contribute to the long-wave instability. We shall isolate several of these effects in four simple models of thin liquid films. These models are constructed to contain the minimum amount of physics necessary to demonstrate the mechanism of the instability for each effect. In § 2, we shall discuss the classical case of an isothermal liquid film flowing down an inclined plane. Two fluids of different densities flowing in a concentric manner in a vertical pipe are examined in § 3. In § 4, we consider an inclined liquid film cooled from below. Finally, a horizontal liquid layer heated along the interface and whose motion is driven by thermocapillary forces is discussed in § 5.

2. Isothermal inclined liquid film. The first and simplest model we shall consider is that of an isothermal liquid film on a rigid plate inclined at the angle β with respect to the horizontal as shown in figure 1. The stability of this film flow was first analyzed by Benjamin (1957) and Yih (1963).

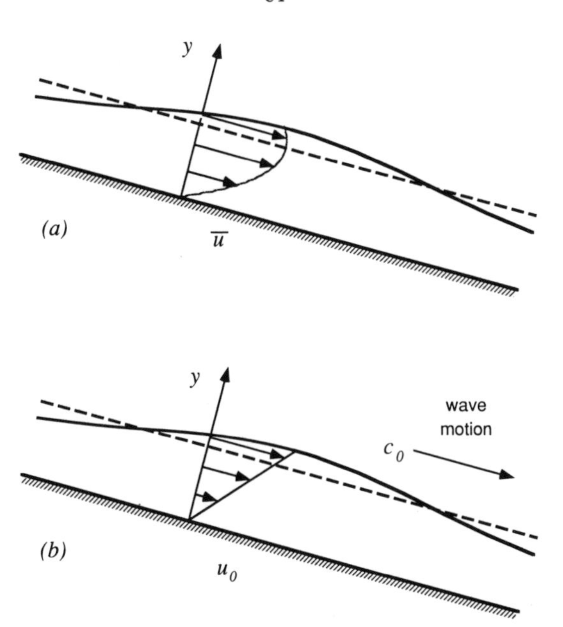

FIG. 2. *Flows in the layer with a long-wave disturbance to the interface. The dashed line is the undeformed free-surface position. (a) The basic-state flow in the layer. The basic-state shear stress at the deformed interfacial position is nonzero. (b) The leading-order flow that balances the nonzero shear stress in the basic-state flow. The resulting kinematic wave motion is downstream.*

The motion in the film is governed by the Navier-Stokes and continuity equations in the bulk liquid, the no-slip boundary condition on the rigid plate, and the kinematic condition and the tangential- and normal-stress balances on the free surface. Our scaling for this problem gives the Reynolds number as $R = \sin(\beta) g d^3/\nu^2$, where ν is the kinematic viscosity of the liquid.

The basic-state flow in the layer in a steady, parallel shear flow of uniform thickness and a longitudinal velocity of $\overline{u} = y - y^2/2$, as shown in figure 2a.

The stability of this flow is studied using standard linear stability theory and normal modes. The wavenumber of the disturbance is α, and the eigenvalue $c = c_r + ic_i$ contains the phase speed c_r and the growth rate αc_i of the disturbance. The system is unstable when c_i is positive.

To examine the response of the system to disturbances with long wavelengths, we use a regular perturbation analysis of the normal-mode disturbance equations for small α. For example, the expansion for the longitudinal normal-mode velocity is

$$\hat{u} = u_0 + \alpha u_1 + \cdots. \tag{1}$$

The result of this perturbation analysis is the eigenvalue

$$c_r = 1 + O(\alpha^2) \tag{2}$$

$$c_i = \alpha \left\{ \frac{2}{15} R - \frac{1}{3} \cot(\beta) \right\} + O(\alpha^2). \tag{3}$$

For neutral stability, $c_i = 0$ and we find

$$R_c = \frac{5}{2} \cot(\beta). \tag{4}$$

This is the result of Benjamin (1957) and Yih (1963) when the difference in scaling is considered.

In c_i, the term $2R/15$ represents the destabilizing effect of inertia, and the term $-\cot(\beta)/3$ represents the stabilizing effect of gravity normal to the layer. These results answer the overall stability question for this system, but they do not provide an understanding of the dynamics of the inertial and gravitational mechanisms that produce these effects. To do this, we need to examine the flow structure in the liquid film in more detail. Recently, three papers were published that presented four different ways to describe these dynamics. Kelly, Goussis, Lin, & Hsu (1989) presented two of these four. Their first description is a global mechanical energy balance for the unstable linear mode. This technique identifies the processes through which energy is fed into the unstable motion from the basic-state flow. Their second description uses a global vorticity balance for the unstable linear mode. This technique shows how disturbance vorticity is generated and destroyed within the layer. For more details of these calculations, see Kelly, et al. (1989).

These two descriptions do provide useful information about the instability. However, one must first calculate the unstable linear mode in order to compute these two global balances. Given a new flow situation, it may not be easy to use the results of a previous calculation to estimate whether the new flow will be stable or not. This is especially true for the vorticity balance since it is difficult to tell how disturbance vorticity will be generated at the lower solid boundary.

Hsieh (1990) discussed the mechanism of the long-wave instability for the flow down an inclined plane using an inviscid-flow argument. His analysis allowed him to calculate an extremely accurate estimate for the point of neutral stability in this flow. However, this does not imply, as he suggests, that the mechanism for the instability is based on inviscid fluid dynamics. The unstable flow is still highly viscous. This paradox of an inviscid analysis giving good estimates for an unstable viscous flow has been resolved by Smith (1991).

The last description of the mechanism for the long-wave instability of the flow down an inclined plane was proposed by Smith (1990a). We shall now briefly describe this mechanism.

In figure 2a, we see the basic-state velocity field in the inclined film and a long-wave disturbance to the interface. At the crest of this disturbance, the basic-state velocity gradient is negative. Since the interface is a stress-free surface, a perturbation shear stress driving a leading-order disturbance flow must be present so that the total stress on the interface is zero. This disturbance longitudinal velocity u_0, shown in figure 2b, is a linear viscous flow whose flow rate increases as the interfacial deformation increases. This motion pushes fluid under the right side of the disturbance crest causing the interface to deflect upwards, and removes fluid from under the left side of the disturbance crest causing the interface to deflect downwards. This kind of interfacial motion is exactly that of a neutral wave propagating to the right at the phase speed $c_r = c_0 = 1$.

The instability of this wave motion is governed by the $O(\alpha)$ longitudinal motion u_1 described by the following boundary value problem from the perturbation analysis

$$\mathrm{D}^2 u_1 - i\,\{p_0 + R\,[(\overline{u} - c_0)u_0 + \overline{u}'(-iv_1)]\} = 0 \tag{5}$$

$$u_1(0) = \mathrm{D}u_1(1) = 0. \tag{6}$$

Here, D indicates a derivative with respect to the variable y.

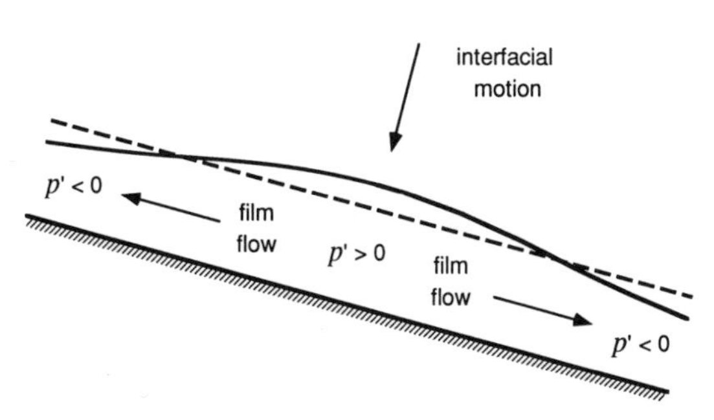

FIG. 3. *The stabilizing film flow that develops in response to a pressure disturbance p' in phase with the interfacial deformation. The dashed line is the undeformed free-surface position.*

Consider a similar viscous flow problem in which u_1 is driven by a pressure perturbation in the layer

$$\mathrm{D}^2 u_1 - ip = 0, \quad u_1(0) = \mathrm{D}u_1(1) = 0. \tag{7}$$

Here, p is the normal-mode pressure, and when $p > 0$ the pressure perturbation p' in the layer is in phase with the interfacial motion. This pressure drives a longitudinal flow in the liquid that is away from the crest of the disturbance as shown in figure 3. The resulting flow is stabilizing since it removes fluid from underneath the crest causing it to decrease in amplitude.

In equation (5), the terms in the curly brackets are all known functions from earlier problems in the perturbation analysis. We can treat this entire collection of terms as a driving stress for the $O(\alpha)$ viscous flow in the layer. The term p_0 is the extra hydrostatic pressure in the layer due to the increased interfacial deflection and the normal component of gravity. It is positive and stabilizing since it tends to push fluid away from the crest of the disturbance. The terms multiplied by the Reynolds number represent the effects of inertia associated with interactions between the basic-state flow and the leading-order disturbance flow. In this problem the sum of these terms is strictly negative for all values of y, although this is not in general true for other problems. The negative inertial terms are destabilizing since they force fluid underneath the disturbance crest. To estimate the critical point of neutral stability, we define the average inertial stress as

$$\mathcal{P} = R \int_0^1 \{(\overline{u} - c_0)u_0 + \overline{u}'(-iv_1)\}\, dy, \tag{8}$$

and solve the viscous problem

$$\mathrm{D}^2 u_1 - i\{p_0 + \mathcal{P}\} = 0, \quad u_1(0) = \mathrm{D}u_1(1) = 0. \tag{9}$$

If $p_0 + \mathcal{P} = 0$, then $u_1 = 0$ and we would have neutral stability at this order. This translates to the neutral stability condition $R_c = 3\,\cot(\beta)$. This is 20% higher than the exact value of $R_c = (5/2)\,\cot(\beta)$.

The inertial terms from equation (5) are the result of advection between the basic-state velocity and the leading-order disturbance flow. Their effect on the viscous flow in the layer is shown in figure 4. If we move with the interfacial disturbance, then the flow in the layer $u_0 - c_0 < 0$ is always to the left. Since the longitudinal motion u_0 is

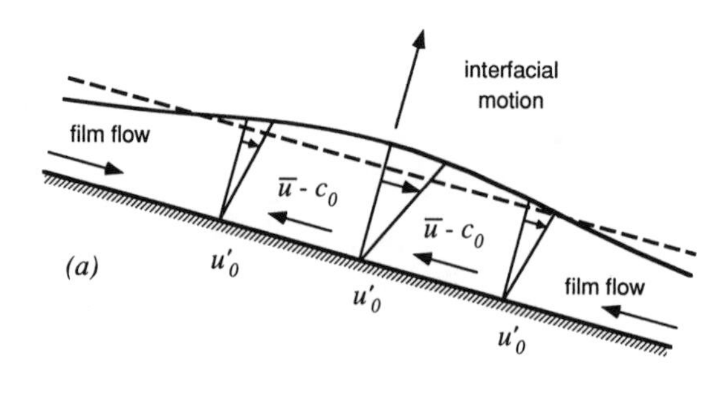

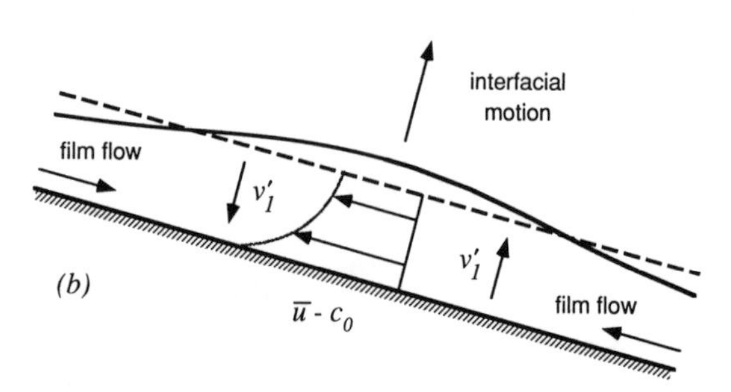

FIG. 4. *Flows in the layer with a long-wave disturbance to the interface. The dashed line is the undeformed free-surface position. (a) A diagram of how the longitudinal advection of momentum within the layer produces a negative inertial stress and a destabilizing film flow. (b) A diagram of how the normal advection of momentum within the layer produces a negative inertial stress and a destabilizing film flow.*

always positive, the inertial stress due to longitudinal advection is $R(\overline{u}-c_0)u_0 < 0$ and therefore destabilizing. In figure 4a, we see that on the right side of the disturbance crest the motion $u_0 - c_0$ moves fluid with a low velocity u_0 over towards regions where the velocity is larger. This inertial effect causes the fluid to slow down, or equivalently, it produces a flow perturbation towards the disturbance crest. Likewise on the left side of the crest, the flow $u_0 - c_0$ moves fluid with a larger u_0 over towards regions where the flow is slower. This effect speeds up the flow on the left side of the crest, which is again a flow perturbation toward the crest. Both of these flow perturbations push fluid underneath the disturbance crest causing the interfacial deflection to increase.

The inertial stress due to normal advection $R\overline{u}'(-iv_1)$ is also negative throughout the entire liquid layer. Its destabilizing effect on the interface is shown in figure 4b. As we move with the disturbance, the normal velocity in the layer is upwards on the right side of the disturbance crest. This motion moves fluid with a larger basic-state velocity upwards into a region where the velocity is lower. This inertial effect speeds up the flow on the right side towards the crest. On the left side of the crest, the disturbance velocity is downwards and the advection slows down the fluid moving to the left, which is equivalent to a flow perturbation toward the crest. The net effect of these motions is to move fluid underneath the crest and to increase the amplitude of the disturbance.

A more global description of this long-wave instability uses the net longitudinal

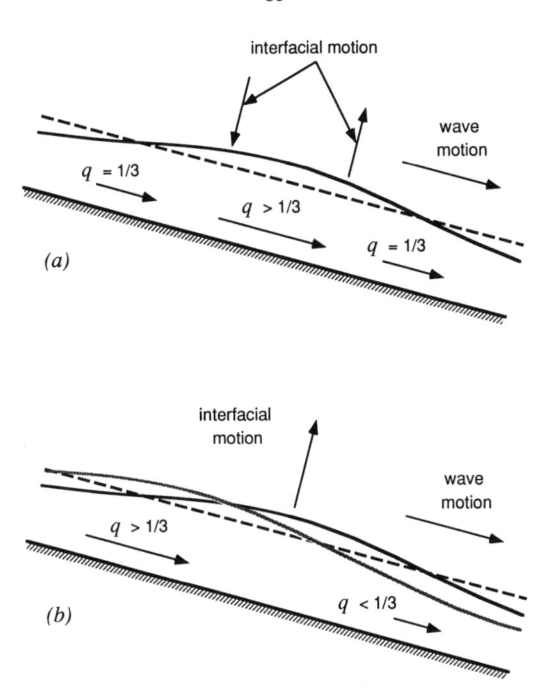

FIG. 5. *Flows in the layer with a long-wave disturbance to the interface. The dashed line is the undeformed free-surface position. (a) The total longitudinal flow and its variation with the interfacial deformation and the production of the neutral kinematic wave motion. (b) The effect of the inertial stress in the layer is to create a phase lag of the longitudinal flow (gray line) with respect to the interfacial deformation. The resulting destabilizing flow brings more fluid underneath the crest which increases the interfacial deformation.*

flow in the layer $q = \eta^3/3 + O(\alpha)$, where η is the local depth of the layer. Note that the flow rate increases with the depth of the layer. Thus, for a long-wave disturbance to the interface, as shown in figure 5a, the flow rate must increase underneath the disturbance crests. This flow distribution pushes fluid under the right side of the disturbance crest and drains fluid from the left side of the disturbance crest. This in turn causes the interfacial deformation of a neutral wave propagating to the right as we saw before. The dominant inertial effect in this system is caused by the unsteadiness of this neutral wave motion. As the right side of a disturbance passes over a point in the layer, the layer depth increases. The flow rate should increase to match $q = \eta^3/3$, but the inertia of the fluid prevents this and so the flow rate is a little bit less than expected. Likewise, on the left side of the disturbance, the flow must decelerate as the disturbance passes. Again, inertia prevents the full deceleration of the flow and so the flow rate is a little bit higher than expected. Figure 5b shows that this inertial effect has produced a small phase lag of the longitudinal flow rate with respect to the interfacial deformation. A longitudinal flow rate that is larger on the left side of the disturbance crest than on the right side will cause an increase in the interfacial deformation. If this motion is larger than the stabilizing motion caused by the normal component of gravity represented by the hydrostatic pressure term, then the system will be unstable.

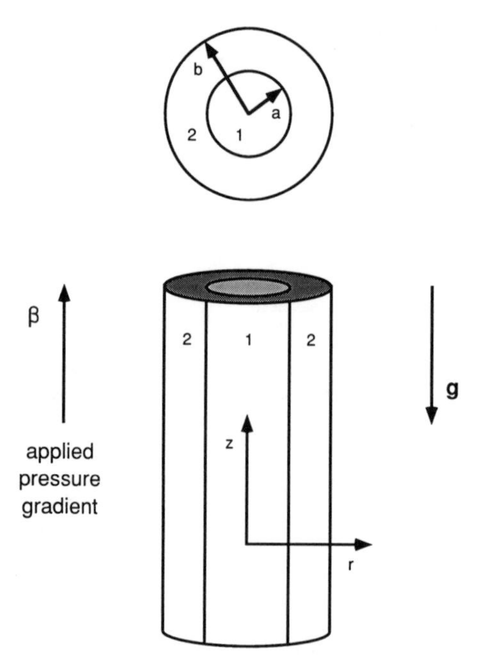

FIG. 6. *The geometry of the flow of two concentric fluids in a pipe.*

3. Concentric two-fluid flow in a vertical pipe. Our second model is intended to examine the effect of an enclosed system on the long-wave interfacial instability. Consider the flow of two fluids in a vertical pipe as shown in figure 6. The fluids are arranged in a concentric manner with fluid 1, of radius a, at the center of the pipe, and fluid 2 in the annular region between fluid 1 and the pipe wall at radius b. Gravity and an applied pressure gradient β act on the fluids. The two fluids have different densities, but all other fluid properties are identical. The governing equations for this flow are the Navier-Stokes equations and continuity, with the no-slip boundary condition on the pipe wall. The interfacial boundary conditions are continuity of velocity, the kinematic condition, and the normal- and tangential-stress balances. After scaling this problem, the relevant dimensionless groups are the Reynolds number, $R = g\triangle\rho a^3/\mu\nu_1$, the pressure gradient $\beta^* = \beta/g\triangle\rho$, the density ratio, $l = \rho_1/\rho_2$, and the radius ratio, $b^* = b/a$, with $\triangle\rho = \rho_1 - \rho_2$. We shall ignore surface tension effects in this discussion. For more details, see Smith (1989).

The basic state for this system is a cylindrical interface between fluids one and two and an axial velocity in each fluid driven by gravity and the applied pressure gradient. The results of a linear stability analysis for this problem are quite complicated. First, since the pipe is vertical, there is no stabilizing effect due to gravity. The stability or instability of the system is determined solely by the inertial stresses that we previously identified for the inclined-film problem. Thus, the stability results are independent of the Reynolds number. Second, the domains of instability for this pipe flow are

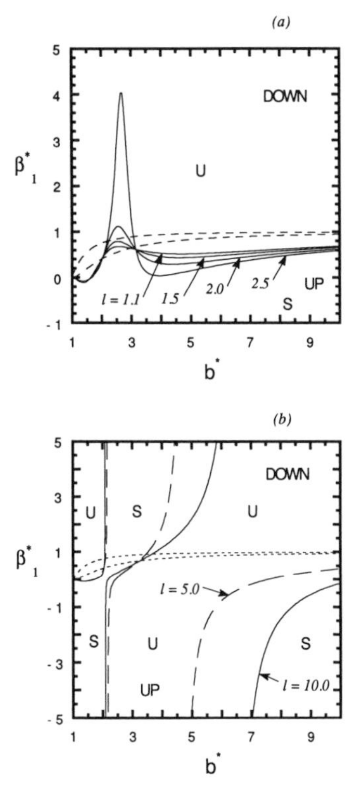

FIG. 7. *The neutral stability curves of the total driving force for motion in the inner fluid β_1^* versus the radius ratio b^* for various values of the density ratio l. Stable regions are marked by the letter S and unstable regions by the letter U. Above the upper short-dashed line the flow in the pipe is all downwards, and below the lower short-dashed line the flow is all upwards. (a) l = 1.1, 1.5, 2.0, and 2.5. (b) l = 5.0 and 10.0.*

a complicated function of β_1^*, the net driving force in fluid 1 (gravity plus pressure gradient), the density ratio of the two fluids l, and the geometry of the flow b^*. Some selected stability curves for $l > 1$ are shown in figure 7a. Here, we see the growth of a large peak at $b^* \approx 2.6$ as l increases from a value of 1.1. Eventually, the smoothly connected neutral curve breaks into three separate pieces for large enough values of b^* as shown in figure 7b. This complicated behavior is a result of the enclosed nature of the flow.

Consider a long-wave disturbance to the interface in this pipe-flow problem. The leading-order response of the system is to form a perturbation shear stress on the interface so that the net tangential stress is continuous across the interface. This stress is proportional to the difference in curvature of the basic-state velocities, and is always downwards. A simple flow field that one may expect to find in the two fluids is shown in figure 8a. The interfacial velocity increases with the interfacial deflection just as we saw in the simple inclined-film problem. In the outer fluid, this motion causes the interface below the crest to move radially inward. The downward motion in the inner fluid causes the interface below the crest to move radially outward. Since the interfacial velocity must be the same for each fluid this situation is impossible.

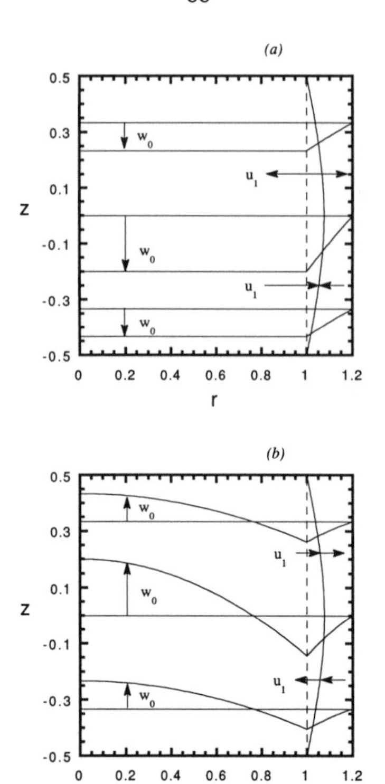

FIG. 8. *The $O(1)$ axial flow w_0 in the pipe in response to a long-wave disturbance to the interface. The dashed line is the undeformed free surface position. The arrows on the deformed interface indicate the magnitude and direction of the radial interfacial velocity u_1 produced by the axial flows in each liquid respectively. The flow (a) without lubrication pressure, and (b) with lubrication pressure.*

The actual behavior for this flow is dictated by the downward motion in the thin layer of fluid on the pipe wall. As this flow pushes fluid below the disturbance crest, the interface moves inward and squeezes the interior fluid. A large pressure gradient forms because the outer fluid pushes against the rigid pipe walls. This pressure drives the inner fluid upward and ensures continuity of radial velocity at the interface. We call this pressure a lubrication pressure in analogy to that seen in a lubrication problem since it is of $O(\alpha^{-1})$. The actual leading-order disturbance axial flow is shown in figure 8b.

The effect of the lubrication pressure on the leading-order axial flow in each fluid depends on the cross-sectional area of the fluid. Figure 9 shows how this velocity changes as the radius ratio of the two fluids change. The behavior of the system is always determined by the interfacially-forced motion of the smaller layer.

The key problem for the determination of stability in this flow is the $O(\alpha)$ axial velocity problem, exactly analogous to the inclined-film model. This problem is

$$\mathrm{D}^2 w_1^{(j)} - i p_0^{(j)} - iRl^{1-j}\left\{(\overline{w}^{(j)} - c_0) w_0^{(j)} + \overline{w}_r^{(j)}(-i u_1^{(j)})\right\} = 0 \tag{10}$$

$$w_1^{(1)}(0) < \infty, \quad w_1^{(2)}(b^*) = 0 \tag{11}$$

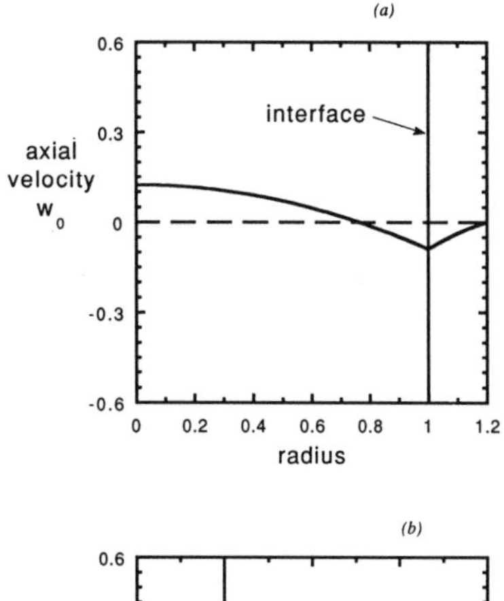

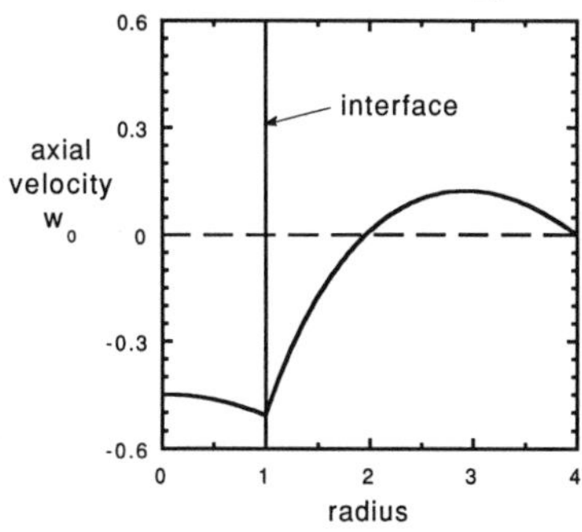

FIG. 9. *The effect of differing liquid thicknesses on the $O(1)$ axial flow in the pipe.*

$$w_1^{(1)}(1) = w_1^{(2)}(1), \quad \mathrm{D}w_1^{(1)}(1) = \mathrm{D}w_1^{(2)}(1). \tag{12}$$

Here, the axial velocity is driven by inertial stresses, while the pressure $p_0^{(j)}$ ensures continuity of radial velocity on the interface at this order.

The average inertial stress in each liquid is defined as

$$\mathcal{P}_j = \frac{R\,l^{1-j}}{A_j} \int_{A_j} \left\{(\overline{w}^{(j)} - c_0)w_0^{(j)} + \overline{w}_r^{(j)}(-iu_1^{(j)})\right\} dA_j. \tag{13}$$

The solution of the simpler differential equation $\mathrm{D}^2 w_1^{(j)} - i\left\{p_0^{(j)} + \mathcal{P}_j\right\} = 0$, with the appropriate boundary conditions, indicates instability for the flow when $\mathcal{P}_2 - \mathcal{P}_1 > 0$. Figure 10 shows how these inertial stresses produce such an unstable flow situation.

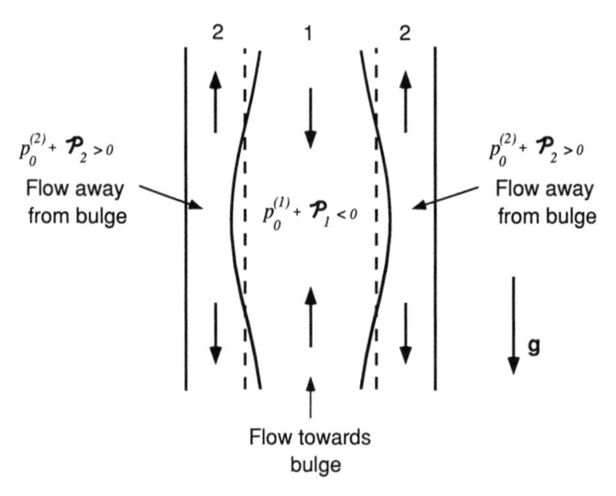

FIG. 10. *The inertial stresses in the pipe and the resulting destabilizing axial flow.*

The inertial stresses $\mathcal{P}_1/R$ and $\mathcal{P}_2/R$ for $\beta_1^* = 1$ are plotted in figure 11. One can predict from this plot all of the features of the neutral stability curves shown in figure 7, including the peak near $b^* = 2.6$ and the appearance of several different stability regions as shown in figure 7b. The complicated behavior of the instability is a result of the sign change of the inertial stresses, which in turn is a result of the change in the leading-order axial velocity due to the lubrication pressure. Thus, we emphasize the importance of the lubrication pressure in understanding the complicated behavior of the long-wave instability in these confined-flow geometries.

4. Inclined liquid film cooled from below. Our third model is an inclined liquid film cooled from below. A sketch of the geometry is shown in figure 12. The rigid plate is held at the fixed temperature T_b^* and the passive gas above the liquid is held at the temperature $T_a^* > T_b^*$. The governing equations for this system are the Navier-Stokes equations, in which we assume that the density is a linear function of the temperature and we do not employ the Boussinesq approximation, the energy equation, and the full conservation of mass equation

$$\rho_t + u\rho_x + v\rho_y + \rho\left\{u_x + v_y\right\} = 0, \tag{14}$$

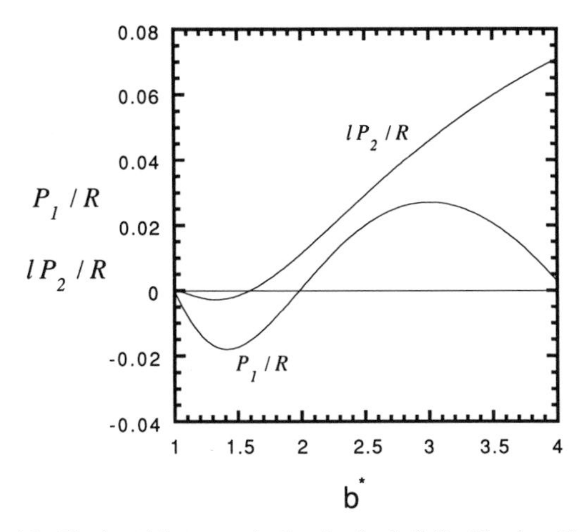

FIG. 11. *The inertial stresses in the pipe for both liquids when* $\beta_1^* = 1$.

where the non-dimensional form of the density is $\rho = 1 - ET$. The boundary conditions for the velocity are the same as for the isothermal film, and here, we add the thermal conditions of a fixed temperature on the rigid plate and a convective heat balance on the free surface.

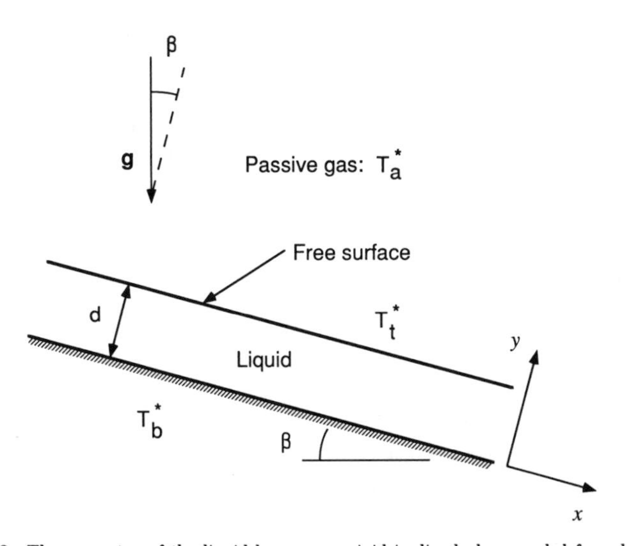

FIG. 12. *The geometry of the liquid layer on a rigid inclined plane cooled from below.*

The important dimensionless parameters are the Reynolds number as for the isothermal film, the Peclet number $Pe = RPr$, the Prandtl number $Pr = \nu_b/\kappa$, the Biot number $B = h_g d/k$, and the expansion number $E = \gamma \triangle T$. Here, ν_b is the kinematic diffusivity referred to the temperature of the bottom surface, κ is the thermal diffusivity, h_g is the free-surface conductance, k is the thermal conductivity, γ is the thermal expansion coefficient of the liquid, and $\triangle T = T_t^* - T_b^* > 0$ for cooling from below. In ΔT, T_i^* is the basic-state temperature of the free surface, which depends on T_a^* and the Biot number. This model for the case of both cooling and heating from below is analyzed and discussed in detail by Smith (1990b).

The result of a long-wave linear stability analysis in which we ignore $O(E)$ terms, but keep $O(E\,Pe)$ terms is

$$c_r = 1 + O(\alpha^2) \tag{15}$$

$$c_i = \alpha\left\{\frac{2}{15}R - \frac{1}{3}\cot(\beta) + E\,Pe\left(\frac{17}{1260} - \frac{73}{2240}\mathcal{B} + \frac{11}{720}\mathcal{B}^2\right)\right\} + O(\alpha^2). \tag{16}$$

Here, $\mathcal{B} \equiv B/(B+1)$ lies between zero and one by definition. The inertial and gravitational terms are exactly the same as for the isothermal layer indicating that the associated physical mechanisms are also the same. For $\mathcal{B} = 1$, the thermal term is negative indicating a stabilizing effect. This is the expected behavior when the layer is cooled from below. However, for $\mathcal{B} = 0$, the thermal term is positive indicating a destabilizing effect. We can explain these results using the following thermal mechanisms.

In an undisturbed liquid layer, the temperature is determined by conduction and is given by the gray line in figure 13a. The *leading-order* temperature field for the disturbed layer is also given by conduction. Consider the case when $\mathcal{B} = 1$, so that the free surface of the layer is isothermal. As a long-wave disturbance to the interfacial position propagates downstream, the temperature in the bulk of the layer must decrease as the interfacial position increases because the temperature on the interface must remain the same. This bulk temperature decrease follows the interfacial deformation and is greatest at the crest of the disturbance as shown in figure 13a. However, the liquid cannot change its temperature as fast as the interfacial deformation changes because of the finite heat capacity of the liquid. Likewise, as the interface position decreases the bulk liquid can not heat up as fast as the interfacial deformation changes. The resulting temperature field, shown in figure 13b, is warmer on the right side of the disturbance crest when compared to pure conduction, and cooler on the left side of the crest. With this temperature distribution, the liquid becomes lighter on the right and heavier on the left. This difference in density creates a buoyancy-driven film flow that is toward the crest of the disturbance from both sides. Thus, the buoyancy effect is destabilizing for this case of an isothermal free surface.

Now consider the effects of direct contraction of the liquid. In the case shown in figure 13b, as the disturbance moves by a point in the liquid, the fluid experiences a temperature drop from the warmer to the colder region. As this occurs, the fluid underneath the disturbance crest contracts. The contraction reduces the volume of the fluid under the crest, causing the disturbance amplitude to decrease. This stabilizing effect is larger than the destabilizing effect of buoyancy. Thus, thermal effects have a net stabilizing influence on a liquid film cooled from below when the Biot number is large.

Now consider the case where we have a fixed heat flux through the free surface of the layer, so that $\mathcal{B} = 0$. In figure 14a, we see that as the free-surface position is deformed there is no effect on the temperature field up to leading order. This is because the leading-order thermal field is conduction dominated and since the heat flux out of the layer at the free surface is fixed and equal to the basic-state heat flux the leading-order temperature can only be zero. In this case, advection of the thermal field by the normal velocity is the important process. In figure 14a, we see that the

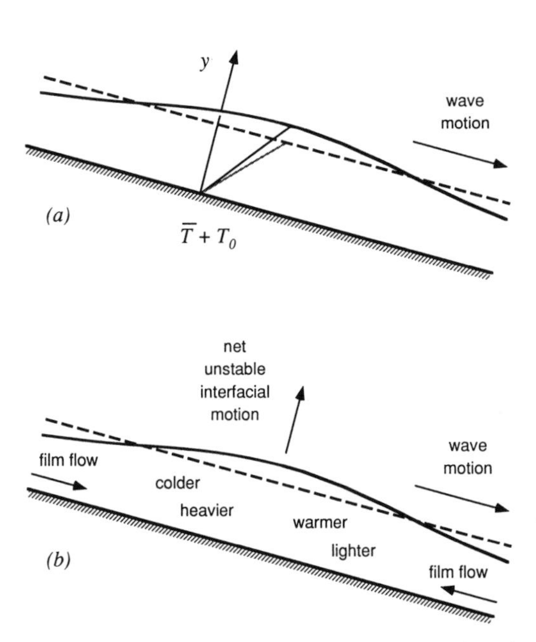

FIG. 13. *The flows and temperature field in the layer with a long-wave disturbance to the interface. The dashed line is the undeformed free-surface position. The interface is an isothermal surface. (a) The gray line is the undisturbed temperature and the solid line is the disturbed temperature in the layer. The bulk liquid has become cooler. (b) The advection of heat due to the wave motion of the disturbance creates the temperature disturbances shown. Buoyancy then produces a destabilizing film flow as shown, but liquid contraction overcomes this flow and stabilizes the layer.*

normal velocity in the layer is upwards on the right side of the crest and downwards on the left. Thus, on the right, the normal velocity will bring cold fluid upwards and cause the fluid on the right side to be relatively colder than expected. Likewise on the left, the downward flow will heat up the fluid a small amount. From the resulting temperature perturbations, shown in figure 14b, we see that buoyancy will cause a stabilizing film flow and that direct liquid expansion will be destabilizing. In this case, the expansion effect dominates and the layer will be unstable as predicted.

It is interesting to note that in this last case of a fixed surface heat flux the system is unstable when we cool the liquid film from below. This counter-intuitive result is due to the effects of direct liquid expansion that are not normally considered in problems of this kind.

5. Horizontal thermocapillary liquid film. The last problem we shall consider is a horizontal liquid film heated along its interface so that a thermocapillary flow is produced. This geometry is shown in figure 15. We insulate the bottom surface and convectively heat the top surface with a dimensional ambient gas temperature of $T_a^* = -bx$. We assume that the surface tension varies linearly with the temperature where γ is the negative rate of change of surface tension with temperature. We shall ignore density changes with temperature. The governing equations are the Navier-Stokes, energy, and continuity equations. In the tangential-stress condition on the interface we balance the viscous stress with the surface-tension gradient.

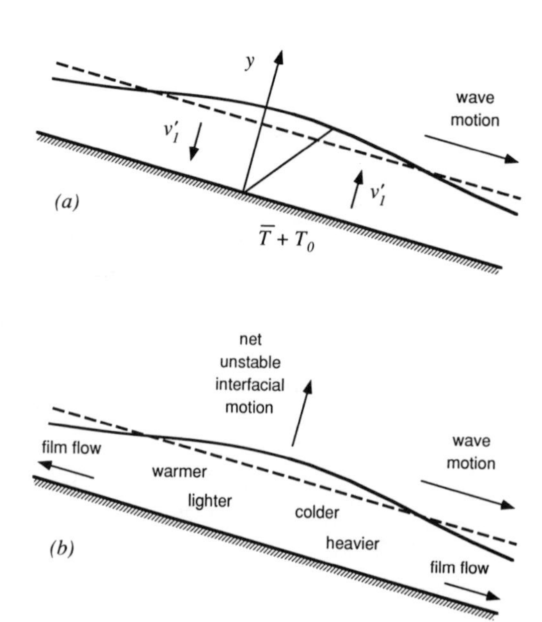

FIG. 14. *The flows and temperature field in the layer with a long-wave disturbance to the interface. The dashed line is the undeformed free-surface position. The heat flux through the interface is fixed. (a) The solid line is the temperature in the layer. It is undisturbed by the interfacial deformation. The normal velocity due to the wave motion of the interface is shown by the arrows. (b) The advection of heat due to the wave motion of the disturbance creates the temperature disturbances shown. Buoyancy then produces a stabilizing film flow as shown, but liquid expansion overcomes this flow and destabilizes the layer.*

The dimensionless groups that appear in this analysis are the Reynolds number $R = \gamma b d^2/\mu\nu$, the Marangoni number $M = \gamma b d^2/\mu\kappa$, the Biot number $B = h_g d/k$, and the gravity number $G = \rho g d/\gamma b$, where μ and ρ are the dynamic viscosity and density of the liquid respectively. See Smith & Davis (1983) for more details of this model when $G = 0$.

In the basic-state velocity field, shown in figure 16a, the surface flow is driven by the thermocapillary shear stress at the interface, and a pressure gradient along the film drives the return flow at the bottom. The net flow across the layer is zero.

The result of the long-wave analysis is

$$c_r = -1/2 + O(\alpha^2) \tag{17}$$

$$c_i = \alpha\left\{\frac{1}{10}R - \frac{1}{3}G + \frac{M}{4B}\right\} + O(\alpha^2). \tag{18}$$

Here, we see that inertia is destabilizing, gravity is stabilizing, and that thermal effects are destabilizing.

The thermal effect is due to thermocapillary motion in the film. In figure 16a we see a long-wave disturbance to the layer and an exaggerated view of how the velocity profile changes with height. The net longitudinal flow in the *disturbed* layer is to the left. This flow moves in the basic-state temperature field that is colder to the right. Thus, the disturbed flow moves colder fluid to the right and cools the fluid under the

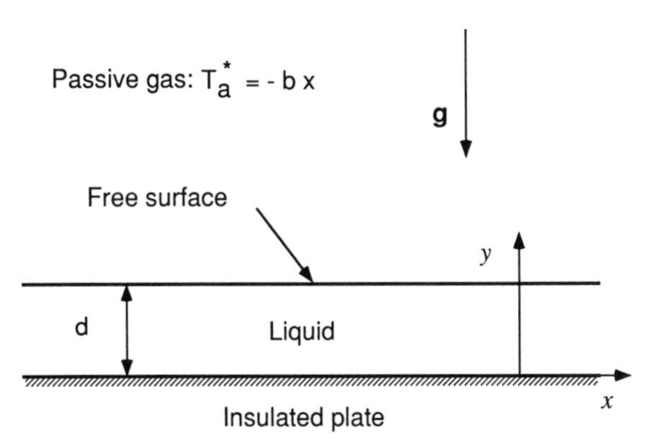

FIG. 15. *The geometry of a horizontal layer driven by thermocapillarity. The layer is heated convectively along the free surface by the passive gas.*

crest. The resulting temperature perturbation is shown in figure 16b; cooler under the crest and warmer under the troughs. This temperature perturbation drives a thermocapillary flow from the troughs to the crest that forces fluid underneath the crest. Thus, thermocapillarity is destabilizing.

6. Conclusions. The four models of thin liquid film flows explored in this paper have shown how long-wave motions in the liquid are produced by gravity normal to the layer, the inertia of the liquid, the confining nature of pipe or channel walls,

buoyancy, direct liquid expansion, and thermocapillarity. The use of a linear stability analysis allowed us to investigate how each of these effects interact with the interfacial deformation to either stabilize or destabilize the layer.

The long-wave instability mechanisms we have described show a great similarity in their fundamental components. The underlying flow in the layer is a parallel, viscous flow driven by the forces applied to the bulk liquid or to the free surface of the layer. For an interfacial disturbance that is much larger than the depth of the liquid, these flows vary as the depth of the liquid changes, but retain their locally parallel characteristic. The temperature profile is locally conduction-dominated when the layer is cooled or heated from below, but is determined by a balance between conduction and advection when the layer is heated horizontally. To leading order, the longitudinal flow and the temperature are either in phase or 180° out of phase with the interfacial deformation. As the disturbance propagates along the interface, gravity normal to the film and the advection of momentum or of heat within the film cause a phase shift in the longitudinal flow. This is accomplished through the action of liquid inertia or through thermally-activated forces such as buoyancy or thermocapillarity. The sense of the phase shift in the longitudinal flow determines the direction of the net flow underneath an interfacial crest. If the net flow is toward the crest and large enough to overcome the effect of liquid thermal contraction, the disturbance will continue to increase in magnitude and the film is unstable.

We have shown that the average inertial stress is a good estimate for the effects of inertia on the instability. And surprisingly, we have found that the thermal expansion of the liquid dominates the buoyancy forces in a film cooled or heated from below. It is this effect that allows a film cooled from below to become unstable when the heat transfer from the free surface is fixed.

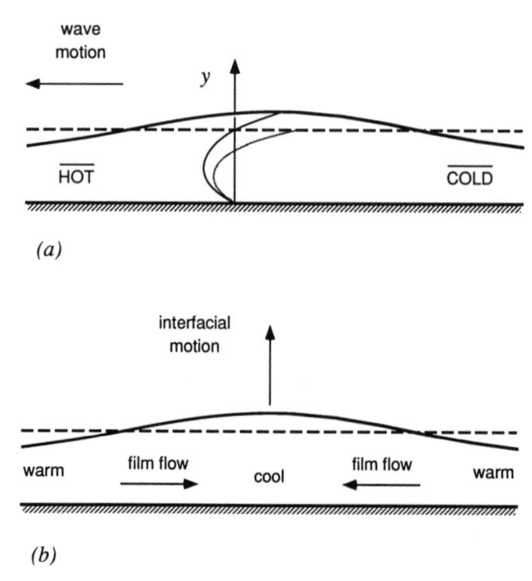

FIG. 16. *Flows in the layer with a long-wave disturbance to the interface. The dashed line is the undeformed free-surface position. (a) The basic-state flow for the undisturbed layer (gray line) and the disturbed layer (solid line). The net longitudinal flow is to the left under an interfacial elevation forcing the wave motion of the disturbance to the left. The basic-state temperature increases to the left. (b) The temperature perturbations due to the cooling by the disturbed longitudinal flow as it moves cooler fluid from the right to the left. Thermocapillarity produces a destabilizing film flow.*

In confined-flow systems with two or more fluids in a pipe or a channel, the rigid walls help produce a large lubrication pressure in the liquids. When the interface is deformed, the thinner liquid layer squeezes the thicker layer by pushing against the rigid wall of the pipe or channel. The resulting lubrication pressure produces large changes in the viscous flow in the system that in turn can produce large changes in the inertial stresses in the liquids. This causes the complicated behavior of these systems that is seen as the thicknesses of the liquid layers change.

The physical mechanisms we have described in this work show that the underlying physics of the long-wave instability in thin liquid films is the same in a great variety of different physical situations. We can use these ideas to help discuss the dozens of research papers published on this instability. In addition, the separate mechanisms associated with the different forces involved in these systems can be combined to understand the response of larger and more complicated physical models. Hopefully, the general understanding that results from this kind of work will be useful to the scientists and engineers who are actively involved in working with these systems.

Finally, we must remember that the long-wave analysis predicts a critical wavenumber of zero, which means an infinitely long disturbance. This, of course, will never be seen in any finite system. Nevertheless, the physics described by the asymptotic analysis is quite robust so we expect the underlying instability mechanisms are at least approximately correct for the small, but finite wavenumbers seen in actual physical systems. More detailed studies at these non-zero wavenumbers are needed to verify the extent of this claim for any specific system.

Acknowledgements. This work was supported by the National Science Foundation, Grant No. MSM–8451093.

REFERENCES

ATHERTON, R. W. & HOMSY, G. M., (1976) Chem. Eng. Comm. **2**, 57.
BENJAMIN, T. B., (1957) J. Fluid Mech. **2**, 554.
BENNEY, D. J., (1966) J. Math. & Phys. **45**, 150.
HSIEH, D. Y., (1990) Phys. Fluids A **2**, 1145.
KELLY, R. E., GOUSSIS, D. A., LIN, S. P., AND HSU, F. K., (1989) Phys. Fluids A **1**, 819.
SMITH, M. K., (1989) Phys. Fluids A **1**, 494.
SMITH, M. K., (1990a) J. Fluid Mech. **217**, 469.
SMITH, M. K., (1990b) J. Fluid Mech. **217**, 469.
SMITH, M. K., (1991) Phys. Fluids A, submitted.
SMITH, M. K. & DAVIS, S. H., (1983) J. Fluid Mech. **121**, 187 .
YIH, C. -S., (1963) Phys. Fluids **6**, 321.

CONVECTIVE AND MORPHOLOGICAL STABILITY DURING DIRECTIONAL SOLIDIFICATION OF THE SUCCINONITRILE-ACETONE SYSTEM

S.R. CORIELL*, B.T. MURRAY*, G.B. McFADDEN*, AND K. LEONARTZ†

Abstract. Convective and interfacial instabilities during directional solidification are considered for a binary system where the coupling of the two modes of instability leads to oscillatory behavior very near the onset of instability. For a limited range of the control parameters, an oscillatory critical mode of instability is actually obtained. The directional solidification model assumes vertical growth of a binary alloy at constant velocity. Buoyant thermosolutal convection and morphological stability are treated via a stability analysis of the linearized governing equations and boundary conditions, which include the Boussinesq form of the Navier-Stokes equations for viscous flow and the required conservation laws for mass and energy in the two phases and at the solid-liquid interface. Numerical results for the stability criteria are obtained using two independent solution procedures. Detailed results are presented for the region of parameters where oscillatory behavior is obtained at or close to onset.

1. Introduction. Crystal growth from a liquid phase or solidification of an alloy combines the nonlinearities associated with fluid dynamics and free boundary problems. Even in the absence of fluid flow, the crystal-melt interface is subject to morphological instability [1], resulting in cellular and dendritic growth. The length scale of the morphological instability is very small compared to the length scale associated with fluid flow, which is typically the container size, i.e., a few centimeters. This disparity in length scales makes realistic numerical studies of the interaction of fluid flow with the crystal-melt interface intractable at present. Brown [2] has reviewed many aspects of efforts to model crystal growth ranging from calculations of heat transfer in the sample container to microscopic cellular interfaces. Progress on understanding the coupling between hydrodynamics and the crystal-melt free boundary has been based on linear and weakly nonlinear stability analyses of idealized geometries and flows [3, 4].

The solidification of a binary alloy at constant velocity, V, with a planar crystal-melt interface in a quiescent fluid provides a base state whose stability can be determined; it is also a reasonable approximation to directional solidification experiments in which a sample is translated at constant velocity in an imposed temperature gradient supplied by a differentially-heated furnace. After initial transients, the temperature and concentration fields are independent of time in a frame moving with the crystal-melt interface, and depend only on distance from the planar interface. We consider solidification in the vertical direction so that in an idealized furnace with no radial (horizontal) gradients of temperature a quiescent state is possible. The temperature increases with distance from the interface in the melt and decreases with distance from the interface in the crystal. For most fluids the density decreases with temperature so that for growth vertically upwards in a pure fluid the density decreases with height; this is a convectively stable arrangement. For growth vertically downwards, heavier fluid is above lighter fluid and there is the possibility of a convective instability.

The concentration of solute changes discontinuously at the crystal-melt interface since the equilibrium concentrations at a given temperature are different in the crystal

* National Institute of Standards and Technology, Gaithersburg, MD 20899 USA.

† ACCESS e.V., D-5100 Aachen, Germany.

and melt. This rejection or preferential incorporation of solute leads to an exponential concentration gradient in the melt with a decay distance D_L/V, where D_L is the solute diffusivity. Since the density of the liquid depends on the solute concentration, the solute field can give rise to heavier liquid above lighter liquid, resulting in the possibility of convective instability. Whether the solute field is convectively stabilizing or destabilizing depends on the growth direction, the sign of density change with respect to concentration, and whether solute is rejected or preferentially incorporated.

Since the density depends on both temperature and concentration, double-diffusive effects are important [5]. For example, it is well known that instability may occur even if the base state is such that the density decreases with height. The situation in which one field is stabilizing and one destabilizing is interesting; if the faster diffusing species (temperature in the solidification problem) is destabilizing the onset of convective instability is usually oscillatory in time. If the slower diffusing species is destabilizing the onset of convective instability is not oscillatory in time. In the simplest double-diffusive problem the temperature and solute gradients are constant, while in the solidification problem the solute gradient is exponential and the temperature gradient is approximately constant.

In addition to convective instability, morphological instability may occur during solidification of a binary alloy. The solute gradient promotes morphological instability while the temperature gradient and the crystal-melt surface tension play a stabilizing role. At low solidification velocities, the role of surface tension is not important in determining the conditions for instability, although it is crucial for determining the wavelength at the onset of instability. Instability occurs when the ratio of the solute and temperature gradients is sufficiently large.

The growth vertically upwards of an alloy that rejects a lighter solute has been extensively studied [6, 7, 8, 9, 10, 11, 12, 13, 14, 15, 16, 17]. The temperature gradient stabilizes both convective and morphological modes while the concentration gradient destabilizes both modes. Viscosity and surface tension also stabilize the convective and morphological modes, respectively. As the concentration gradient increases, instability will occur but whether it is convective, morphological, or a coupled instability depends on the solidification velocity and temperature gradient. In the absence of coupling between the morphological and convective mode, we would anticipate that the onset of instability would be non-oscillatory in time, that is, the principle of exchange of stabilities would hold. This has been proven for the morphological mode under fairly general conditions [18] and is certainly true for double diffusive convection with linear gradients in this regime. Many of the studies considered the growth of lead-tin alloys vertically upwards with temperature gradients in the melt of the order of 100 K/cm. For given concentration and temperature gradients, the onset of instability was by a convective mode at slow solidification velocities changing to a morphological mode as the velocity increased. The wavelength associated with the morphological mode was considerably smaller than that associated with the convective mode. At velocities near the changeover from convective to morphological modes of instability, modes that were oscillatory in time were calculated, but these were not the most dangerous mode. Thus, in the lead-tin system at typical temperature gradients, the onset of instability was not oscillatory in time. When calculations were carried out for the succinonitrile-ethanol system, there was a narrow range of velocities for which the onset of instability was oscillatory in time [8, 10]. From their studies of the coupling between morphological and convective modes, Caroli et al [11] suggested that an oscillatory onset would occur in the lead-tin system at very low temperature gradients; numerical calculations with $G_L = 0.1$ K/cm by one the authors verified this prediction [3]. Jenkins [16] has recently studied the onset of oscillatory modes in the lead-tin system as well. Riley and Davis [17] have undertaken a systematic study of the possibility of oscillatory modes associated with strong coupling between convective and morphological instabilities.

Rex and Sahm [19] have investigated convective instabilities in aluminum-magnesium alloys during space experiments. Experiments are being planned by Leonartz and colleagues to investigate convective and morphological instabilities in the transparent organic alloy, succinonitrile-acetone. In this system, the lighter acetone is rejected at the solidifying interface. Here, we report the results of linear stability calculations for growth vertically upwards. The properties of succinonitrile-acetone are similar to those of succinonitrile-ethanol [8, 10] where instabilities which are oscillatory in time have been calculated. In the succinonitrile-acetone system, we also find such oscillatory instabilities, and we investigate them in detail here. In addition to the numerical methods used previously [6], we develop a second independent solution technique based on Chebyshev pseudospectral discretization of the governing differential equations which results in a generalized matrix eigenvalue problem for the temporal growth rate of the instabilities.

2. Governing equations. In order to investigate the onset and development of convective and morphological instabilities during directional solidification, the governing equations are written in terms of a Cartesian coordinate system (x, y, z) which is attached to the initially planar crystal-melt interface moving at constant growth velocity V in the positive z direction, as shown in Fig. 1. The liquid region extends from the interface position, $z = 0$, to $z \to \infty$, while the solid region extends to $z \to -\infty$. Assuming an incompressible Newtonian fluid, the basic governing equations for the problem are the conservation of mass equation, the Navier-Stokes equations, the solute conservation equation in the liquid, and the conservation of energy equations in the liquid and solid phases. The Boussinesq approximation is employed to treat buoyant convection, where the density in the liquid phase is assumed to be a linear function of the temperature and solute concentration. The solid-liquid interface is allowed to deform and its position is characterized by $z = h(x, y, t)$. The dimensional governing equations and boundary conditions for the directional solidification model used here are given in Coriell et. al [6]. We summarize the important aspects of the model and present a dimensionless form of the linearized equations and boundary conditions used to obtain the stability results presented here.

The fluid velocity in the melt is described by the vector field $\boldsymbol{u} = (u, v, w)$. This velocity is measured in the laboratory frame in which the crystal is at rest, so that in the undisturbed state $\bar{\boldsymbol{u}} = (0, 0, -\epsilon V)$. Here, the parameter ϵ accounts for bulk flow due to density change upon solidification ($\epsilon = \rho_S/\rho_L - 1$, where ρ_S and ρ_L are the densities of the solid and liquid phases, respectively,) and the overbar is used to represent base state quantities. Under steady-state conditions, the base state solutions for the concentration in the liquid, c, and the temperature fields in the liquid and solid, T_L and T_S, respectively, depend only on the vertical coordinate z and are given by

$$\bar{c} = c_\infty[1 + \frac{1-k}{k}\exp(-\frac{\rho V}{D_L}z)]$$

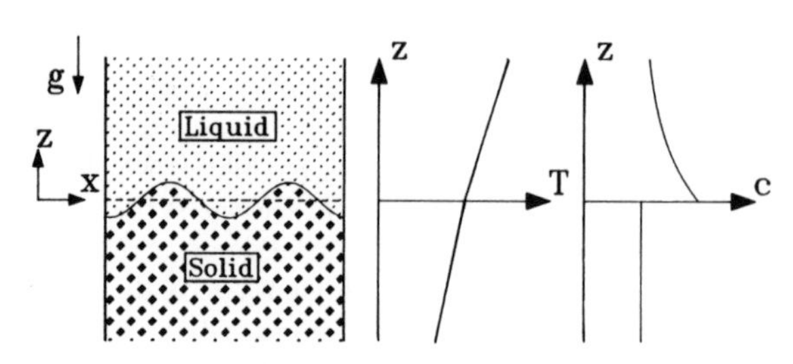

FIG. 1. *Schematic diagram of the directional solidification configuration.*

$$\bar{T}_L = T_m + \frac{mc_\infty}{k} + \frac{\kappa_L G_L}{\rho V}[1 - \exp(-\frac{\rho V}{\kappa_L} z)]$$

$$\bar{T}_S = T_m + \frac{mc_\infty}{k} + \frac{\kappa_S G_S}{V}[1 - \exp(-\frac{V}{\kappa_S} z)]$$

where c_∞ is the bulk concentration, k is the equilibrium segregation coefficient, D_L is the solute diffusion coefficient in the liquid, G_L and G_S are the temperature gradients at the interface in the liquid and solid, respectively, T_m is the melting temperature of pure solvent, m is the slope of the liquidus, κ_S and κ_L are the solid and liquid thermal diffusivities, respectively, and the parameter ρ is the density ratio ($\rho = \rho_S/\rho_L$). In addition, the model equations contain the fluid pressure, p; however, the base state for the pressure is not explicitly given, since it can be eliminated from the linear stability equations.

For the linear stability analysis of the base state, the flow field variables are written as the superposition of the base state component (overbars) and a perturbation (hats). The perturbed quantities are Fourier analyzed in the lateral directions and exponential time-dependence is assumed, so that the flow variables are written as

$$(1) \qquad \begin{pmatrix} u(x,y,z,t) \\ v(x,y,z,t) \\ w(x,y,z,t) \\ p(x,y,z,t) \\ c(x,y,z,t) \\ T_L(x,y,z,t) \\ T_S(x,y,z,t) \\ h(x,y,t) \end{pmatrix} = \begin{pmatrix} 0 \\ 0 \\ -\epsilon V \\ \bar{p}(z) \\ \bar{c}(z) \\ \bar{T}_L(z) \\ \bar{T}_S(z) \\ 0 \end{pmatrix} + \begin{pmatrix} \hat{u}(z) \\ \hat{v}(z) \\ \hat{w}(z) \\ \hat{p}(z) \\ \hat{c}(z) \\ \hat{T}_L(z) \\ \hat{T}_S(z) \\ \hat{h} \end{pmatrix} \exp(\sigma t + ia_x x + ia_y y),$$

where σ is the complex temporal growth rate, and a_x and a_y are the wavenumbers in the lateral directions. Governing equations for the perturbation quantities are obtained by substituting the above quantities into the complete set of nonlinear model equations and linearizing in the perturbation quantities.

For the linear stability analysis, the fluid equations can be reduced to solving a single fourth-order equation for the the perturbed vertical velocity $\hat{w}$, using the standard manipulation for hydrodynamic stability analyses [20]. The variables are made dimensionless by introducing the length scale D_L/V and the time scale D_L/V^2, and by scaling the flow velocity by the solidification velocity V, the solute field by the far field concentration c_∞, and the temperature by $G_L D_L/V$. The dimensionless linearized equations for the stability problem are

$$\text{(2a)} \quad \frac{1}{\mathrm{Sc}}[\sigma(D^2-a^2)\hat{w} - \rho(D^2-a^2)D\hat{w}] = (D^2-a^2)^2\hat{w} - a^2(\frac{\mathrm{Sc}}{\mathrm{Pr}}\mathrm{Ra}\hat{T}_L + \mathrm{Rs}\hat{c}),$$

$$\text{(2b)} \quad \sigma\hat{T}_L - \rho D\hat{T}_L + \exp(-\rho\frac{\mathrm{Pr}}{\mathrm{Sc}}z)\hat{w} = \frac{\mathrm{Sc}}{\mathrm{Pr}}(D^2-a^2)\hat{T}_L,$$

$$\text{(2c)} \quad \sigma\hat{c} - \rho D\hat{c} - \frac{\rho(1-k)}{k}\exp(-\rho z)\hat{w} = (D^2-a^2)\hat{c},$$

$$\text{(2d)} \quad \sigma\hat{T}_S - \hat{T}_S = \kappa\frac{\mathrm{Sc}}{\mathrm{Pr}}(D^2-a^2)\hat{T}_S,$$

where we have defined the operator $D = d/dz$, and a is the magnitude of the lateral wavenumber ($a = \sqrt{a_x^2 + a_y^2}$). Here, $\mathrm{Sc} = \nu/D_L$ is the Schmidt number, $\mathrm{Pr} = \nu/\kappa_L$ is the Prandtl number, $\mathrm{Ra} = g\alpha G_L(D_L/V)^4/(\nu\kappa_L)$ and $\mathrm{Rs} = g\beta c_\infty(D_L/V)^3/(\nu D_L)$ are the thermal and solutal Rayleigh numbers, respectively, and $\kappa = \kappa_S/\kappa_L$ is the thermal diffusivity ratio.

The linearized dimensionless boundary conditions at the interface, $z = 0$, are

$$\text{(3a)} \quad \hat{w} = \sigma(1-\rho)\hat{h},$$

$$\text{(3b)} \quad D\hat{w} = a^2(\rho-1)\hat{h},$$

$$\text{(3c)} \quad \frac{1}{\rho}D\hat{c} + (1-k)\hat{c} + \rho(1-k)\hat{h} = \sigma\frac{(k-1)}{k}\hat{h},$$

$$\text{(3d)} \quad -g_L\,D\hat{T}_L + qg_L\,D\hat{T}_S + \frac{\mathrm{Pr}}{\mathrm{Sc}}[\rho g_L - \frac{g_L}{\kappa} - \frac{1}{\kappa}]\hat{h} = \sigma\hat{h},$$

$$\text{(3e)} \quad \hat{T}_L - \hat{T}_S + [1 - \frac{(1+g_L)}{qg_L}]\hat{h} = 0,$$

$$\text{(3f)} \quad \frac{k}{(k-1)}M\hat{c} + \hat{T}_L + [1 + \tilde{\gamma}a^2 - \rho M]\hat{h} = 0,$$

where $g_L = k_L G_L/(L_V V)$ is a dimensionless temperature gradient, $q = k_S/k_L$ is the thermal conductivity ratio, $M = (k-1)mc_\infty V/(kD_L G_L)$ is the Morphological number, and $\tilde{\gamma} = T_m\Gamma V^2/(D_L{}^2 G_L)$ is a capillarity parameter. The various physical parameters appearing in the problem are listed in Table 1, while Table 2 summarizes the dimensionless parameters that appear in the equations and boundary conditions. In addition to the interface conditions, the perturbation quantities are required to decay as $z \to \pm\infty$.

3. Numerical solution. Two distinct solution procedures were used to solve the differential eigenvalue problem described above. The first approach is described in detail in Coriell et. al [6] and is only briefly summarized here. The coupled set of linear two-point boundary value problems is solved using the computer code SUPORT [21], which uses superposition of numerically integrated solutions with an orthonormalization procedure to maintain the linear independence of the solution set. A high-order Adams-type method is used for the numerical integration in the SUPORT code. Because the temperature in the solid, $\hat{T}_S$, is coupled to the variables in the liquid only through the interface boundary conditions, it is possible to solve for $\hat{T}_S$ analytically on the semi-infinite solid domain. Thus, only the eighth-order system of equations in the liquid region are integrated numerically. For numerical purposes, we assume that the spatial domain in the liquid extends from the interface $z = 0$ to a truncated value of the semi-infinite domain denoted by h_l, where the perturbation quantities are set equal to zero. The magnitude of h_l is taken sufficiently large to make the stability results insensitive to the chosen value.

The homogeneous eigenvalue problem in the first approach is converted into an inhomogeneous, nonsingular problem using the approach suggested by Keller [22], which involves modifying some of the homogeneous boundary conditions in the problem so that nontrivial solutions are obtained. The original boundary conditions are then used as conditions to determine selected eigenvalue parameters via an iteration procedure. Typically, neutral modes are desired, and these are found by setting the real part of the complex growth rate, σ_r equal to zero. In the general case, when the imaginary part of the growth rate, σ_i, is nonzero (oscillatory modes), one parameter (e.g., Rs) and σ_i are used as the iteration parameters. For steady modes ($\sigma_i = 0$), this involves iteration on a single parameter.

The second approach employed consists of discretizing the coupled set of ordinary differential equations using the pseudospectral technique implemented in the physical domain as described in [23]. The approach amounts to expanding the solutions in terms of truncated series of Chebyshev polynomials $T_n(s)$, for example,

$$\hat{w}(z) = \sum_{n=0}^{N} w_n T_n(2z/h_l - 1).$$

The pseudospectral discretization requires that the solution expansions satisfy the governing equations at the specific collocation points for the Chebyshev polynomials. When implemented in the physical domain the unknowns are the solution values at the collocation points. The spatial differential operators in the governing partial differential equations are replaced by discrete matrix operators. Thus, the differential eigenvalue problem is transformed into a matrix eigenvalue problem with the complex growth rate σ as the eigenvalue. In this approach, it is no longer convenient to use the analytic solution for $\hat{T}_S$ and so the equation for $\hat{T}_S$ is discretized along with the equations for the liquid variables. A far-field boundary in the solid is set at the value $z = -h_l$. With $N + 1$ spatial modes per each field variable and a single coefficient for the interface deformation, $\hat{h}$, a generalized eigenvalue problem of size $4N + 5$ is obtained. In this approach, neutral modes are found by using a rootfinding procedure to determine the values of Rs for which $\sigma_r = 0$. The advantage of this approach is that it provides $4N + 5$ approximations to the temporal eigenvalues, which simplifies the characterization of stability boundaries in regions where multiple transition points exist.

4. Results and discussion. For the present model of directional solidification, the formulation is simplified by writing the equations and boundary conditions in dimensionless form, and for a general study of the problem, the results may be more simply characterized in terms of the dimensionless parameters. However, for a specific alloy system, it is convenient to present the results in terms of the physical parameters, c_∞, V, G_L, which are controlled in a directional solidification experiment. Here, results are presented for a single binary alloy consisting of small concentrations of acetone dissolved in succinonitrile. For this system, values for the physical parameters are given in Table I and the dimensionless quantities that are independent of c_∞, V, and G_L are given in Table II; parameter values for the system of lead containing small amounts of tin are included for comparison. The value of the solutal expansion coefficient for succinonitrile-acetone has been estimated from the densities of the pure materials; the remaining values are based on experimental measurements. The density ratio $\rho = 1.03$ for succinonitrile; we have used the value unity in our calculations.

Fig. 2 is a stability diagram, where the critical acetone concentration for instability (the minimum of c_∞ for all wavenumbers) is plotted versus the solidification velocity, V, for a fixed value of the temperature gradient, $G_L = 20$ K/cm. The upper solid curve is the morphological mode of instability, for which the critical concentration value decreases with increasing V. For V greater than about 5 μm/s, the morphological mode is the minimum critical value. At low values of V, the convective mode of instability occurs for the lowest concentration values, which is represented by the lower solid curve in the figure; in this region we also show the morphological mode even though it is not the most dangerous. Both the convective and morphological modes represented by the two solid curves correspond to steady onset ($\sigma_i = 0$). Oscillatory onset is obtained for a small range of solidification velocities (4.1 to 4.8 μm/s) in the region where the two modes of instability occur for approximately the same c_∞ value. This region of oscillatory onset is represented by the dashed curve. There is a jump discontinuity in the critical concentration at the end of the convective branch between the steady and oscillatory modes near $V = 4$ μm/s.

The next five figures give a detailed picture of the behavior in the region near the discontinuity and where oscillatory onset is obtained. In Figs. 3-5, we plot acetone concentration, c_∞, as a function of the disturbance wavenumber, $\omega = Va/D_L$; also shown is the value of σ_i for the oscillatory mode of instability. In Fig. 3, results are shown for $G_L = 20$ K/ cm and $V = 4.02$ μm/s. Solid lines are used to represent steady branches of the neutral curves, while dashed lines represent oscillatory branches. At this value of V, there is an oscillatory branch which occurs at lower values of ω. A small closed steady neutral curve intersects the oscillatory branch close to the minimum point. At larger wavenumbers, there is a steady branch corresponding to the morphological mode of instability. The closed steady neutral curve is the remaining lowest portion of the convective mode which is dominant at lower solidification velocities. Except for this small steady region, the oscillatory mode becomes unstable at lower values of c_∞ than the steady morphological branch that dominates for larger wavenumbers. In Figs. 4 and 5, c_∞ is plotted versus ω for $V = 4.02$ and 4.03 μm/s, respectively. The axes scales have been reduced to concentrate on the region near the closed neutral curve. As V is increased from 4.02 to 4.03 μm/s, the closed region becomes an isola as it detaches from the oscillatory branch. At the same time, the isola shrinks in size as V is increased. Although the results are not shown, as V is increased further the isola continues to shrink which leads to the discontinuity

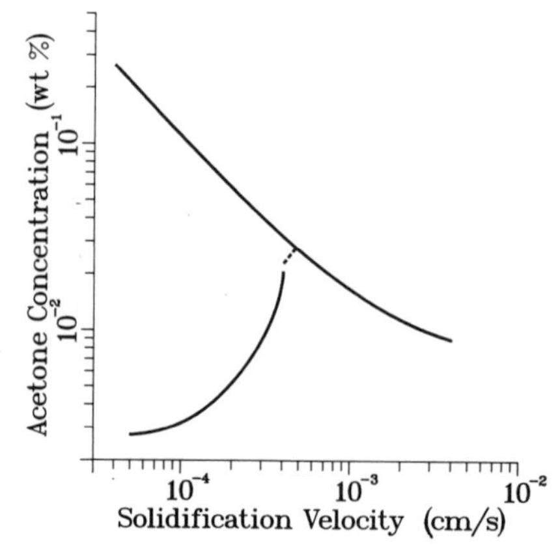

FIG. 2. *The critical acetone concentration at the onset of instability as a function of solidification velocity. The solid curves represent steady onset, while the dashed curve indicates oscillatory onset.*

as the critical concentration jumps to the higher value on the oscillatory branch. For a small range of V beyond the value where the isola disappears, the most unstable mode is oscillatory.

To further understand the behavior in the vicinity of the closed portion of the neutral curve, Figs. 6 and 7 show the real part of the complex growth rate, σ_r, as a function of the acetone concentration, c_∞, for V equal to 4.02 and 4.03 μm/s, respectively. The wavenumber is fixed at the value near the center of the isola shown in Fig. 5, $\omega = 27$ cm^{-1}. In both Figs. 6 and 7, σ_r first becomes positive as a steady mode as indicated by the solid curve. As c_∞ is increased, σ_r reaches a maximum and then decreases back through zero consistent with the closed neutral curve. At the lower velocity, $V = 4.02$ μm/s, σ_r for the oscillatory mode intersects the steady curve for positive values of σ_r. At the higher velocity, $V = 4.03$ μm/s, the curves are shifted downward, such that, the intersection occurs for negative σ_r and leads to the isolated neutral region. As V is increased further, the downward shift in the curves would continue until the maximum of the steady branch passes through the point $\sigma_r = 0$ and the critical concentration jumps to the oscillatory value, accounting for the discontinuity.

More general results for the system are presented in Fig. 8, where c_∞ is plotted as a function of G_L for four values of the solidification velocity ($V = 0.5$, 1, 2 and 4 μm/s). For the three lower velocities, the critical concentration increases with G_L as typically expected, since the temperature field is a stabilizing effect for both the convective and morphological modes of instability. However, for $V = 4$ μm/s there are two modes of instability where the mode changes from oscillatory at lower temperature gradients to steady at higher values. While the critical concentration increases with G_L for the oscillatory mode, the steady mode displays a minimum. Near the intersection point of the oscillatory and steady branches, the critical concentration along the steady branch decreases with increasing temperature gradient. Thus, in a narrow region of parameter space, the temperature field has a destabilizing effect.

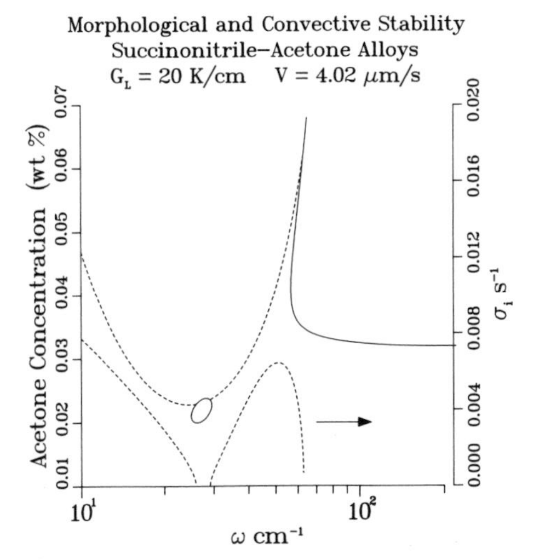

FIG. 3. *The acetone concentration at the onset of instability as a function of wavenumber; the solid curves represent steady onset, while the dashed curve indicates oscillatory onset. The values of the oscillatory frequency, σ_i, are given by axis on the right.*

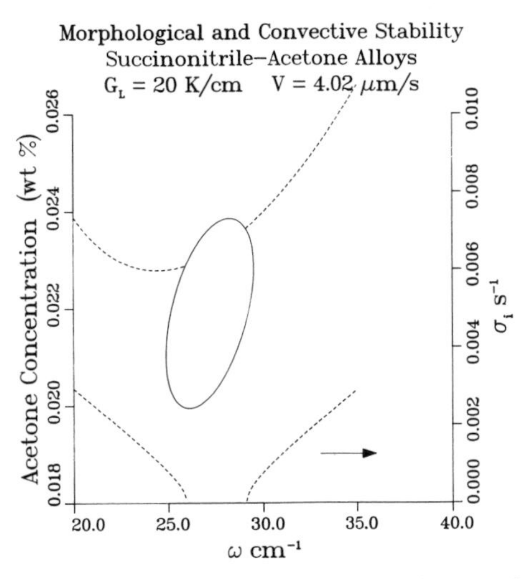

FIG. 4. *The acetone concentration at the onset of instability as a function of wavenumber; the solid curves represent steady onset, while the dashed curve indicates oscillatory onset. The values of the oscillatory frequency, σ_i, are given by the axis on the right.*

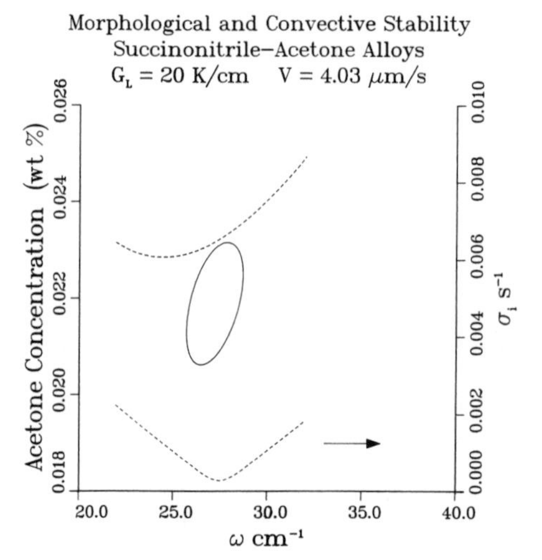

FIG. 5. *The acetone concentration at the onset of instability as a function of wavenumber; the solid curves represent steady onset, while the dashed curve indicates oscillatory onset. The values of the oscillatory frequency, σ_i, are given by the axis on the right.*

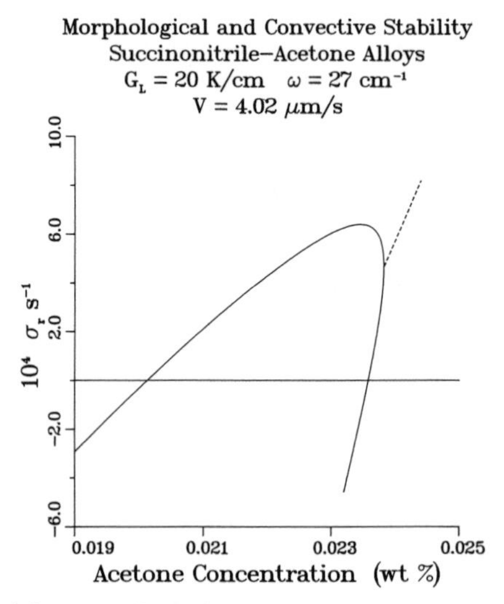

FIG. 6. *The real part of the temporal growth rate, σ_r, as a function of the acetone concentration. The imaginary part, σ_i, is zero for the solid curve and non-zero for the dashed curve.*

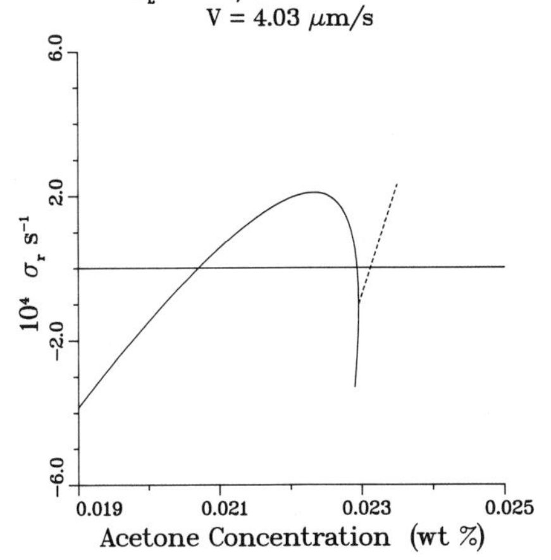

FIG. 7. *The real part of the temporal growth rate, σ_r, as a function of the acetone concentration. The imaginary part, σ_i, is zero for the solid curve and non-zero for the dashed curve.*

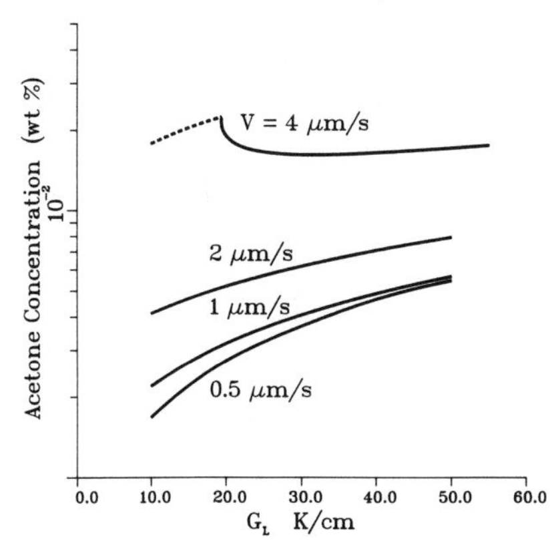

FIG. 8. *The critical acetone concentration at the onset of instability as a function of the temperature gradient in the liquid. The solid curves represent steady onset, while the dashed curve indicates oscillatory onset.*

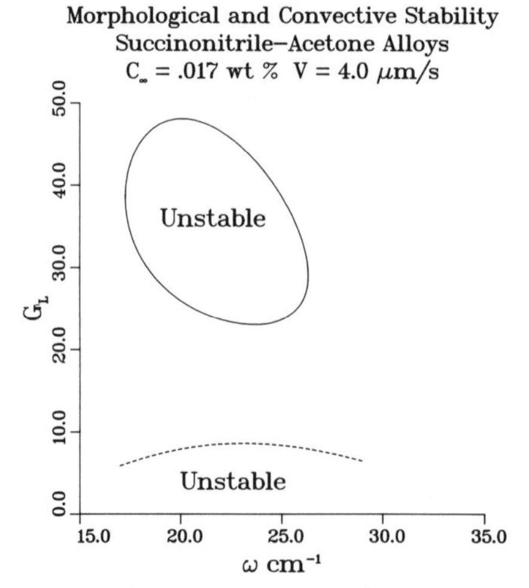

FIG. 9. *The temperature gradient at the stability-instability transition as a function of wavenumber; the solid curve represents steady onset, while the dashed curve indicates oscillatory onset.*

To more fully investigate this behavior, in Fig. 9 we plot neutral values of G_L as a function of wavenumber for fixed concentration and velocity which are in the vicinity of the transition between the oscillatory and steady modes. For very large temperature gradients the system is stable. As the temperature gradient decreases, a steady mode of instability occurs for $G_L = 48$ K/ cm; the system regains stability for $G_L = 23$ K/ cm. As the temperature gradient further decreases, an oscillatory mode of instability occurs for G_L less than 9 K/ cm. The wavenumbers of the various modes are approximately the same. At even smaller values of G_L, morphological instability occurs at large wavenumbers, but this behavior is not included in the figure. The occurrence of the closed region for the steady mode is consistent with the minimum in the steady branch of Fig. 8. Some details of the changes in topology of Fig. 9 with concentration variation can be deduced from Fig. 8. For example, as the concentration decreases, the distance between the closed region and the oscillatory curve increases, while the size of the closed region decreases. A complete description of the changes in topology of these curves requires further investigation.

In conclusion, we have investigated the onset of convective and morphological instabilities in the succinonitrile-acetone system with special emphasis on solidification velocities and temperature gradients for which the onset of instability is oscillatory in time resulting from coupling of the morphological and convective modes of instability. Using a pseudospectral discretization of the governing equations, we have developed a numerical solution procedure which is particularly advantageous for studying oscillatory modes of instability.

5. Acknowledgements. This work was conducted with the support of the Microgravity Science and Applications Division of the National Aeronautics and Space Administration.

Table I

Dimensional Parameters for the succinonitrile-acetone (SCN-ACT) and lead-tin (Pb-Sn) systems

		SCN-ACT	Pb-Sn	
solidification velocity	V	–	–	cm/s
bulk liquid concentration	c_∞	–	–	wt%
temperature gradient	G_L	–	–	K/cm
liquid diffusion coefficient	D_L	1.3×10^{-5}	3.0×10^{-5}	cm^2/s
kinematic viscosity	ν	2.6×10^{-2}	2.43×10^{-3}	cm^2/s
liquid thermal diffusivity	κ_L	1.14×10^{-3}	0.108	cm^2/s
solid thermal diffusivity	κ_S	1.16×10^{-3}	0.202	cm^2/s
liquid thermal conductivity	k_L	2.23×10^{-3}	0.159	J /(cm K s)
solid thermal conductivity	k_S	2.25×10^{-3}	0.297	J /(cm K s)
thermal expansion coefficient	α	1.07×10^{-3}	1.15×10^{-4}	K^{-1}
solutal expansion coefficient	β	2.5×10^{-3}	5.2×10^{-3}	$wt\%^{-1}$
latent heat per unit volume	L_V	47.8	256.0	J/cm^3
liquidus slope	m	-2.8	-2.33	K/wt%
capillarity parameter	$T_m\Gamma$	6.2×10^{-6}	1.0×10^{-5}	cm K

Table II

Dimensionless Parameters for the succinonitrile-acetone (SCN-ACT) and lead-tin (Pb-Sn) systems

		SCN-ACT	Pb-Sn
thermal Rayleigh number	$\mathrm{Ra} = g\alpha G_L(D_L/V)^4/(\nu\kappa_L)$	–	–
solutal Rayleigh number	$\mathrm{Rs} = g\beta c_\infty(D_L/V)^3/(\nu D_L)$	–	–
Morphological number	$M = (k-1)mc_\infty V/(kD_LG_L)$	–	–
temperature gradient	$g_L = k_LG_L/(L_VV)$	–	–
capillarity parameter	$\tilde{\gamma} = T_m\Gamma V^2/({D_L}^2G_L)$	–	–
distribution coefficient	k	0.1	0.3
Prandtl number	$\mathrm{Pr} = \nu/\kappa_L$	22.8	0.0225
Schmidt number	$\mathrm{Sc} = \nu/D_L$	2000	81.0
conductivity ratio	$q = k_S/k_L$	1.009	1.87
diffusivity ratio	$\kappa = \kappa_S/\kappa_L$	1.018	1.87
density ratio	$\rho = \rho_S/\rho_L$	–	–
vertical domain height	$h_l = HV/D_L$	–	–

REFERENCES

[1] W. W. Mullins, and R. F. Sekerka, *Stability of a planar interface during solidification of a dilute binary alloy*, J. Appl. Phys., 35 (1964) pp. 444-451.

[2] R. A. Brown, *Theory of transport processes in single crystal growth from the melt*, AIChE J. 34 (1988) pp. 881-911.

[3] S. H. Davis, *Hydrodynamic interactions in directional solidification*, J. Fluid Mech., 212 (1990) pp. 241-262.

[4] M. E. Glicksman, S. R. Coriell, and G. B. McFadden, *Interaction of flows with the crystal-melt interface*, Ann. Rev. Fluid Mech. 18 (1986) 307-335.

[5] J. S. Turner, *Buoyancy Effects in Fluids*, Cambridge University Press, Cambridge, 1973, pp. 251-287.

[6] S. R. Coriell, M. R. Cordes, W. J. Boettinger, and R. F. Sekerka, *Convective and interfacial instabilities during unidirectional solidification of a binary alloy*, J. Crystal Growth, 49 (1980), pp. 13-28.

[7] D. T. J. Hurle, E. Jakeman, and A. A. Wheeler, *Effect of solutal convection on the morphological stability of a binary alloy*, J. Crystal Growth, 58 (1982) pp. 163-179.

[8] R. J. Schaefer and S. R. Coriell, *Convective and interfacial instabilities during solidification of succinonitrile containing ethanol*, in *Materials Processing in the Reduced Gravity Environment of Space*, G. E. Rindome, ed., Elsevier, Amsterdam, 1982 pp. 479-489.

[9] D. T. J. Hurle, E. Jakeman, and A. A. Wheeler, *Hydrodynamic stability of the melt during solidification of a binary alloy*, Phys. Fluids, 26 (1983) pp. 624-626.

[10] R. J. Schaefer and S. R. Coriell, *Convection-induced distortion of a solid-liquid interface*, Met. Trans. 15A (1984) pp. 2109-2115.

[11] B. Caroli, C. Caroli, C. Misbah, and B. Roulet, *Solutal convection and morphological instability in directional solidification of binary alloys*, J. Phys. (Paris), 46 (1985) pp. 401-413.

[12] B. Caroli, C. Caroli, C. Misbah, and B. Roulet, *Solutal convection and morphological instability in directional solidification of binary alloys. II. Effect of the density difference between the two phases*, J. Phys. (Paris), 46 (1985) pp. 1657-1665.

[13] D. R. Jenkins, *Nonlinear analysis of convective and morphological instability during solidification of a dilute binary alloy*, PhysicoChemical Hydrodynamics, 6 (1985) pp. 521-537.

[14] D. R. Jenkins, *Nonlinear interaction of morphological and convective instabilities during solidification of a dilute binary alloy*, IMA J. Appl. Math., 35 (1985) pp. 145-157.

[15] M. Hennenberg, A. Rouzaud, J. J. Favier, and D. Camel, *Morphological and thermosolutal instabilities inside a deformable solute boundary layer during directional solidification. I. Theoretical methods*, J. Phys. (Paris), 48 (1987) pp. 173-183.

[16] D. R. Jenkins, *Oscillatory instability in a model of directional solidification* J. Crystal Growth, 102 (1990), pp. 481-490.

[17] D. S. Riley and S. H. Davis, *Do the morphological and convective instabilities ever resonate during the directional solidification of a dilute binary mixture*, IMA J. Appl. Math., in press.

[18] S. R. Coriell, G. B. McFadden, P. W. Voorhees, and R. F. Sekerka, *Stability of a planar interface during solidification of a multicomponent system*, J. Crystal Growth 82, (1987) pp. 295-302.

[19] S. Rex and P. R. Sahm, *Planar front solidification of Al-Mg alloy - crystallization front convection*, in *Symposium on Scientific Results of the German Spacelab Mission D1*, P. R. Sahm, R. Jansen and M. H. Keller, eds., DFVLR, Koln, 1987 pp. 222-230.

[20] S. Chandrasekhar, *Hydrodynamic and Hydromagnetic Stability*, Dover, New York, 1981, p. 20.

[21] M.R. Scott and H.A. Watts, *Computational solution of linear two-point boundary value problems via orthonormalization*, SIAM J. Numer. Anal. 14 (1977) pp. 40-70.

[22] H.B. Keller, *Numerical Solutions of Two Point Boundary Value Problems*, Regional Conference Series in Applied Mathematics 24, SIAM, Philadelphia, 1976.

[23] C. Canuto, M. Y. Hussaini, A. Quarteroni, and T. A. Zang, *Spectral Methods in Fluid Mechanics*, Springer, New York, 1988.